T0136716

ENERGY THE GREAT DRIVER

ENERGY THE GREAT DRIVER

SEVEN REVOLUTIONS AND THE CHALLENGES OF CLIMATE CHANGE

R. GARETH WYN JONES

UNIVERSITY OF WALES PRESS
2019

© R. Gareth Wyn Jones, 2019

All rights reserved. No part of this book may be reproduced in any material form (including photocopying or storing it in any medium by electronic means and whether or not transiently or incidentally to some other use of this publication) without the written permission of the copyright owner except in accordance with the provisions of the Copyright, Designs and Patents Act 1988. Applications for the copyright owner's written permission to reproduce any part of this publication should be addressed to the University of Wales Press, University Registry, King Edward VII Avenue, Cardiff CF10 3NS.

www.uwp.co.uk

British Library Cataloguing-in-Publication Data
A catalogue record for this book is available from the British Library.

ISBN 978-1-78683-423-2
eISBN 978-1-78683-424-9

The right of R. Gareth Wyn Jones to be identified as author of this work has been asserted by him in accordance with sections 77 and 79 of the Copyright, Designs and Patents Act 1988.

The University of Wales Press gratefully acknowledges the financial support of the Books Council of Wales and of the Learned Society of Wales in publication of this work.

Typeset by Marie Doherty
Printed by CPI Antony Rowe, Melksham

Contents

Acknowledgements

I am very grateful to a number of friends and colleagues for reading and commenting on parts of the manuscript as it evolved, including John Llywelyn Williams, Roger Leigh, Tony Rippin, Timm Hoffman, John Raven, Barrie Johnson, James Intriligator and Duncan Brown. In doing so they saved me from a number of errors, but, of course, any that remain lie at my own door.

I am indebted to my son, Huw, and grandsons, Euros and Aled, and to Andrew Packwood for their help with the diagrams. I am also grateful to Professor Gareth Ffowc Roberts, Sir John Meurig Thomas and Sir John Houghton for encouraging me to turn the original concept into this text.

I am even more deeply indebted to my wife for her patience and support with a project that should not have taken up so much of my time at this stage in my life.

List of Illustrations

Figures

Tables

Prologue

The ruins of Aleppo (Halap/Halabi) are a reminder of the fragility of peace and civilised behaviour. Over the years I have had the good fortune to work with and for ICARDA (International Center for Agricultural Research in the Dry Areas) in nearby Tel Hadya and to spend time in Aleppo itself enjoying one of the oldest and most diverse cities on earth. Lying in the Fertile Crescent roughly equidistant between the Euphrates and the eastern Mediterranean, it has been a vibrant commercial and cultural centre for thousands of years. It was home to a bewildering number of ethnic groups and religious affiliations, not only Muslim, such as Sunni, Shia, Alawite, Ishmaeli and various Sufi sects, but five bishops guiding their various Christian flocks – some speaking Aramaic, the language of Jesus. Over the millennia it has been ruled by Akkadians, Hittites, Persians, Armenians, Romans, Byzantines, Arab Sassanids and Umayyads, the Egyptian Mamluks and the French, amongst others. Although I was well aware of the dangers of crossing the Baathist regime, the city appeared very safe, largely at ease with itself and to be gradually liberalising. Until the US interventions in Kuwait and Iraq, I saw few women wearing veils or headscarves and the Armenian restaurants in the old city were well patronised. Hopefully the city will rise from the calamity of the current civil war as it has overcome and absorbed past invasions and disasters. But the ruins should be a fearful warning of the thin line separating us from barbarity.

The war was started by internal divisions within Syria, but it has been aggravated and perpetuated by regional and other foreign powers pursuing their own interests and vendettas through proxies. However, another fact has also worried me. In the years leading to the civil war, according to NASA, Syria suffered the worst drought in 400 years. I do not know how much this contributed to the uprising but the susceptibility of the livelihoods of many farmers to drought, their dependence on pumping water from retreating groundwater aquifers and importance of the allocation of water rights were plain to see. The war rumbles on, and the fear and despair that have gripped Syria are palpable. As climate change and global warming tighten their grip on populous countries, many already buffeted by a range of complex and intractable problems, one must ask whether this is a harbinger of a more stormy and violent future. What will be the fate of drought-stricken Iran or parts of the Indian subcontinent where farmers are being compelled to cope not only with debts but with temperatures in the high forties and humidities so high that they override the human body's temperature controls? Or the thousands seeking to escape to poverty and environmental degradation, but are largely unwelcome in Europe and the USA? As I start to write, Cape Town has about a month's supply of drinking water left and it is still high summer. How will these scarce water resources be allocated if the rains do not come? Will some groups be protected and others not? Will communal violence erupt?

This volume illustrates my personal journey from plant biochemistry and cell biology to, initially, a growing interest in agricultural research for development. Later it led to work on sustainability in various southern African rangeland communities – some very close to the climatic cliff ledge. These EU-funded projects involved specialists from many nations and a wide range of disciplines. They required economists, social scientists, hydrologists, agriculturalists and ecologists to collaborate. Hopefully we helped the target communities; certainly we, as collaborators, learnt much from each other. It became apparent that neither the mitigation of nor adaption to anthropogenic climate change should be filed under 'environmental issues' or 'low-carbon energy technologies'. The challenges lie in the totality of our conventional social, political and economic constructs. People flourish or languish, live or die within the constraints imposed by those systems.

Well before James Lovelock and Lynn Margulis popularised the Gaia hypothesis of a planetary 'geophysiology', in the late 1950s my old professor in Bangor, W. Charles Evans FRS, was preaching the essential continuity of the geospheres and biospheres. He conceived chemical threads leading from geology and geochemistry through soils and their abundant microbiology and the various atmospheric inputs, to higher plants and animals, be the latter humans or ruminants. Being strongly influenced by his vision, my own work has led me to perceive further threads leading from cell biology to human societies including communities as different as the subsistence farmers or the herders of the Maloti mountains in Lesotho and the residual hunter-gatherers of the Matsheng in the central Kgalagadi (Kalahari) or the inhabitants of the bustling megapolises of Karachi and London in our globalised world.

Two chains stood out. The first was energy and the consequential ability to do and to control work and to generate power. The second was homeostatic regulation. The former is of course also fundamental to the climate change issue as it is now being expressed as a dangerous manifestation of the age-old relationship between energy flow and biology – a relationship that lies at the heart of all planetary life. The latter is, however, far less well known.

I became aware of the concept after, inspired by postgraduate lectures by the late Professor Jack Dainty on plant biophysics at University of California, Los Angeles, I started working on plant nutrition and stress adaptation. This led to an appreciation of the importance of homeostatic regulation in plants, especially a requirement for the integration of regulatory responses in the metabolically active cytoplasm and volumetrically dominant but metabolically less dynamic vacuole. Although the cytoplasm was, as we said, biophysically and biochemically 'fastidious', the vacuole made the most of whatever solutes were available and cheap in a given ecological niche to balance out the water and biochemical relations of plant cells [1][2][3]. This dichotomy and its regulatory implications will inform some of the concepts considered later.

From this limited beginning and many years of reading, particularly the works of Antonio Damasio, Nick Lane, Suzana Herculano-Houzel, Richard Wrangham, Daniel Kahneman and recently Vaclav Smil, I became increasingly convinced that energy availability and flow provides an underlying coherence to much of biological and geological evolution. I also realised that the acquisition of power, that is the amount of energy used per unit, has led to an acceleration in the development of biological and social complexity and that homeostasis supplies the essential mechanism to stabilise this growing complexity. Furthermore, I have become persuaded that the concept of homeostasis is as relevant to our understanding the energy-dependent social, material, cultural and economic constructs of *Homo sapiens* as it is to cell biology.

This short volume attempts to synthesise and develop the concepts and hypotheses proposed by many other authors. I gladly acknowledge my debt to them for their original work, while hoping that my integration and augmentation of these ideas adds to our understanding. In my judgement, the picture that emerges has important implications both for our understanding of the current energy crisis and for how humans may or may not respond to the challenge of climate change and the future flourishing, or otherwise, of humanity and the rest of the natural world.

CHAPTER I

Introduction

The interrelated hypotheses proposed in this volume have a simple basis. It is that energy enables work and power, which, in turn, lead inexorably to ordered complexity. New sources of energy or step changes in energy use have catalysed greater complexity, be it in pre-human biological systems and their structures or in human society and our constructs. Such structures, derived from and dependent on this flow of energy, would be very short-lived unless sustained by specific stabilising mechanisms. In the case of cells and organisms, these ensure that core internal functions are maintained, if possible well maintained, and flourish in the face of external change and internal demands. Consequently, as biological and latterly human complexity has evolved, these stabilising systems must also evolve and develop to accommodate each new challenge. These simple concepts, drawing on well-known physical and biological principles, can, I believe, be extended to provide us with useful insights into human society. As I explore later in this book, they have, in turn, profound implications for our modern dilemmas, including how we react to the multiple challenges of global warming and climate change and the Anthropocene era.

In chapters II to VI I outline the six major energy step changes that have come to define our planet's bio-, hydro- and geo-spheres over the last 4 billion years. In chapters VIII and IX I explore the evolution of the stabilising homeostatic mechanisms and other emergent properties arising from these energy revolutions. In chapter X I summarise the background to the seventh revolution, this is responding to anthropogenic climate change. This is the revolution in which we are now embroiled but whose outcome is profoundly uncertain. The final three chapters are devoted to an exploration of this uncertainty and the implications of the whole hypothesis to our modern society, including the social and economic constructs that have given a significant proportion of humanity unprecedented material affluence. However, let us begin with the physics.

In classical physics *energy* is defined as the ability to do *work* but both the sources of the energy and the work carried out may vary widely. The science of thermodynamics is about the quantitative interchangeability of the forms of energy and the ensuing work. It is one of the great intellectual achievements of science. The first law of thermodynamics tells us that the energy cannot be created or destroyed in closed systems and the second law that, again in closed systems, entropy, which can be viewed as an approximation to molecular disorder, can never decrease over time, i.e. over time a closed system like our Universe tends towards higher entropy or greater disorder. But in open systems such as those that characterise living organisms, one component, e.g. a cell or organism, can accumulate low entropy or order at the expense of increasing the entropy, crudely the molecular disorder, of its external environment. If the source of energy is inhibited, the ordered structure will gradually revert to increasing disorder or higher entropy unless further work is done on that system (see note 1 for a very brief discussion of the entropy). As explained succinctly by Mackay [4] and Smil [5], talking of 'energy being used' is inexact, although it is part of normal speech and easily understood. It would be more appropriate and correct to refer to energy being transformed from one form to another with no overall loss. Nevertheless, such energy may initially be in a form that can be exploited, e.g. as heat or forms of chemical energy, while moving from a low to a high entropic state. During these transformations work can be done, although in every case some heat energy is dissipated. For ease of understanding I will continue to write of 'energy being used'.

These concepts are best appreciated by practical examples. Harnessing the gravity-induced flow of water, which contains kinetic energy dependent mainly on the height of the fall, will turn, mechanically, a generator, which by exploiting electromagnetic phenomena, can produce an electrical current. In such a current, electrons will flow down a gradient of potential electrical energy and this energy may, in turn, be used to light an electric bulb (emitting heat and light energy) or run another motor (kinetic energy). The latter occurs in the growing trend to electric cars (EVs), and, of course, our computer drives. At each step, an energy transition is exploited and, while a small amount of heat is lost, useful work can be accomplished. Alternatively we may consider the potential chemical energy within a hydrocarbon, be it a sugar or petrol. This can be released by the controlled 'burning' (oxidation) of that hydrocarbon to allow animals and humans to live and work and the latter to run most of their cars and trucks. As an aside, it is worth noting that the heat lost during the energy metabolism of animals, such as cattle, was exploited by our forebears. They built their cowsheds abutting or beneath their dwellings to exploit the 'animal heat' released to combat the cold; a very low cost, albeit primitive, central heating system. A cow, depending on its size and physiological status, may release heat energy equivalent of up to a

1 kW electric fire burning continuously [4]. I have happy memories of in my youth helping a local farmer to hand-milk on cold winter nights and feeling warm and cosy while doing so. On a less cosy note, for millennia the chemical energy embodied in an animal has supplied, on their natural death or slaughter, the energy needs of other animals and microbes in the food chain.

Rarely has the role of energy, in its various guises in the natural and human worlds, been part of normal, everyday conversation. Its universality has bred invisibility. Recently, however, the political economy of energy has commanded greater public attention because of the climate change debate and US President Trump assuming the role of the cheerleader for a highly atmospherically polluting form of chemical energy – coal. In the older generation, there may also be distant memories of the impact of the sudden rise in the oil prices in the 1970s and, certainly in the UK, a concern that so many people live in fuel poverty in dank, cold houses.

Energy and work are critical to all aspects of life on this planet and indeed to non-biological events such as volcanic eruptions, the movement of the earth's tectonic plates and the nuclear fusion reactions in stars, including the Sun. In the biological world, when suitably coupled, the more energy that becomes available, the greater is the potential for structural complexity, specialisation and diversity. As I have noted, in physics *power* is defined as the rate at which *work* is accomplished, i.e. energy use per unit of time. This, I suggest, is equally true in human society. Here an ability to focus and direct work at a particular location at a specific time on a defined task is critical to power. However, it tells us nothing about the direction or utility or morality of the work carried out or indeed how the power has been generated. Nevertheless, much political and economic leverage and power is derived from an ability to control and exploit energy supplies – a theme that will be developed later in this volume.

The concept of energy allowing work and generating power is not new. In reasserting and elaborating on it, I am restating a fundamental relationship recognised for more than a century. The Nobel Laureate Wilhelm Ostwald [6] wrote in 1912: 'free energy is therefore the capital consumed by all creatures of all kinds and by its conversion everything is done'. Ostwald's insight was developed by several other authors, including Howard Odum [7], a pioneer of ecosystems research and the mapping of the flows of energy and matter through them. He suggested that 'the availability of power sources determines the amount of work activity that can exit, and control of these power flows determines the power in man's affairs and in his relative influence on nature'. Vaclav Smil in his recent magisterial volume *Energy and Civilization: A History* [5] reaffirms this basic premise:

> all natural processes and all human actions are, in the most fundamental physical sense, transformations of energy. Civilization's advances can be seen as a quest for the higher energy use required to produce increased food harvests, to mobilise a greater output and variety of materials, to produce more, and more diverse, goods, to enable higher mobility, and to create access to virtually unlimited amounts of information. (p. 385)

In this volume I am expanding on these assertions by claiming that this trail can be traced back to the emergence of life on Earth and forward to the current crisis of climate change and global warming. I will also supplement and expand upon them by exploring other processes, especially that of homeostasis [8] (note 2), which, crucially I argue, mediates and stabilises the relationship between energy, work and complexity.

I highlight six critical step changes in the transformation of energy into material complexity, suggesting that these have led to the overwhelming global dominance of *Homo sapiens* and have shaped our society and indeed the whole of the biosphere. These transformative events, sequentially shown in Table 1, were: (1) the energising of the first living cell(s); (2) the harvesting of the Sun's incoming radiant energy; (3) the evolution of complex, eukaryotic cells with substantially more energy per gene; (4) the investment of addition food energy in the complex and powerful hominid brain; (5) the acquisition of more energy for human society through settled agriculture; and, finally, (6) the use of fossil hydrocarbon fuels to fire the Industrial Revolution. The timeline of these revolutionary step changes and their relationship to some other major events in our planet's history is shown in Figure 1.

Table 1: The major energy events

Revolutions	Approx. time ago	Traits
1. Living cell	~4 billion years (bys)	Prokaryotes: universal proton motive force.
2. Harvesting the Sun	~2.7 bys	Oxygenic Photosynthesis: oxygenation.
3. Complex cells	~2 to 1.7 bys	Eukaryotes: mitochondrial energy: multi-cellularity, specialisation, Darwinian selection.
4. Hominid factor	~2 million years (mys)	Cooking, brain food, mental capacity.
5. More food	10 to 5,000 years	Agriculture, human density and specialists.
6. Fossil fuels	250 years	Steam, electric and internal combustion engines. Science, technology, capitalism, commerce.
7. Current hiatus	Present	Anthropogenic global warming, the Anthropocene era.

Figure 1: Timeline of major events

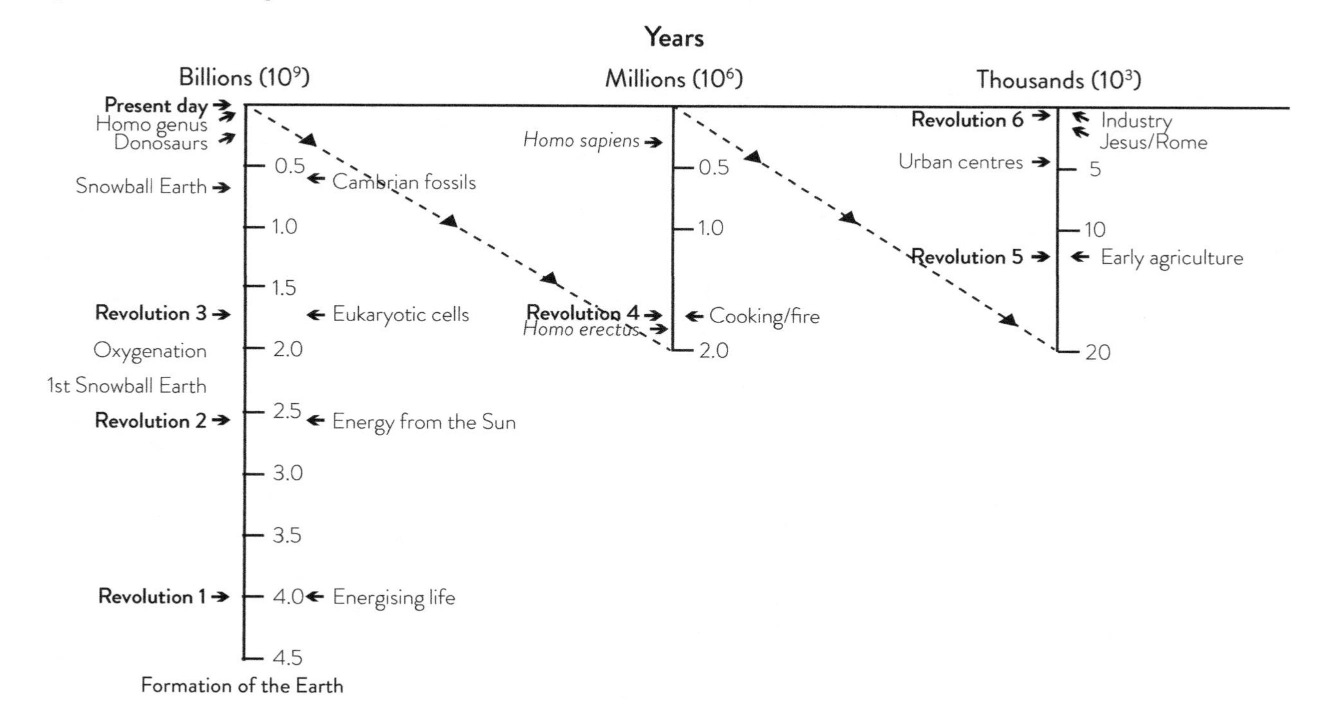

These 'revolutions', especially the early ones, were very slow-burning affairs, stretching over vast periods of time and, in each case, the dominant system of the previous regime retained its biological, geological and social significance. Each 'revolution' was not a single event but complex series of changes and innovations that together made for a step change in the history of our planet. I readily acknowledge that other important but less dramatic changes, many indeed also energy-related, occurred in the intervening periods. However, I contend that these six events in Table 1 and Figure 1 were the planetary game changers.

Conceptualising life on this earth as a series of fundamental revolutions is central to the work of John Maynard Smith and Eörs Szathmáry [9] and, more recently, Tim Lenton and Andrew Watson [10]. These studies embrace somewhat different perspectives. Szathmáry and Maynard Smith propose a list (see Table 2) based on their specialist knowledge and interest in evolutionary genetics; Lenton and Watson from an expertise in geochemistry, earth systems and astrobiology. The latter work owes much to Lovelock's Gaia hypothesis, which, broadly, suggests that Planet Earth is itself

Table 2: Major transitions in evolution

No.	Major transitions
1.	Replicating molecules to populations of molecules in compartments
2.	Unlinked replicators to chromosomes (linked genes)
3.	Ribonucleic acid (RNA) as genes and enzymes prior to deoxyribonucleic acid (DNA) genetic code
4.	Prokaryotes to eukaryotes
5.	Asexual clones to sexual populations
6.	Protists to animals, plants and fungi
7.	Solitary units to colonies (non-reproductive castes)
8.	Primate society to human societies (origin of language)

From Maynard Smith and Szathmáry [9].

a self-regulating system comparable to a biological organism. As might be anticipated, in his own work, Lovelock also emphasises the importance of energy fluxes to earth systems (e.g. [11]). As might be hoped and expected, there are similarities in the crucial events identified in the three hypotheses.

In their excellent book *Revolutions that Made the Earth,* Lenton and Watson [10] highlight four major revolutionary transformations in the history of Earth systems, namely the development of photosynthesis, the emergence of eukaryotes, the agricultural revolution and anthropogenic climate change. These transformations align closely with mine. I hope that my interpretation augments this previous work by focusing on the basic relationship between energy, work and power (this is work per unit time), by including the human social context and, specifically, by exploring the interrelationships between work and complexity and homeostatic regulation at the biological and social levels.

The phenomenon of homeostasis is not widely discussed outside specialised cell biology (see note 2). To try and explain its significance I must explore, briefly, some of the challenges facing a single cell.

Much popular and scientific attention has understandably been focused on the central roles of DNA and molecular genetics in life on Earth and its evolution [12][13][14]. Watson and Crick's discovery of the DNA double helix and the later elucidation of triplet code linking the genetic code to the amino acid sequences in proteins is one of science's proudest achievements (note 3). However, from the very last universal common ancestor (LUCA) to the present (note 4), a reliable and controllable flow of energy, plus, of course, the presence of the relevant chemical building blocks, has been equally essential to life (note 5) (see, for example, [15]).

The internal compartment of a cell, known as the cytoplasm, has a chemical and physical composition demonstrably very different to that of the highly variable external environment. Its composition is remarkably stable and consistent. Specific 'homeostatic' mechanisms then maintain this defined, well-regulated, but dynamic intracellular, aqueous environment well out of thermodynamic equilibrium with the extra cellular environment. Energy must be expended and work done continually to maintain this 'out-of-equilibrium' entity. Indeed, a formidable variety of homeostatic mechanisms are to be found in even the simplest single-cell organisms. They allow such an organism not only to regulate its internal environment but to respond to a great variety of external stimuli. Such external stimuli might have the capacity to influence, for good or ill, the organism's internal integrity and its ability to survive and reproduce. Simple examples would be the need to minimise or avoid negative impacts, such as excessive heat or toxins or dehydration, and, positively, to promote access to potential nutrients or to light or to more benign growth conditions. To anticipate later chapters, one can suggest that cells are capable of seeking conditions that are compatible with their own wellbeing. This regulatory capacity must have arisen in parallel with the beginnings of the first viable cell.

A cell must, for example, be able to maintain its cytoplasm at a pH near neutrality (i.e. close to pH 7) as well as maintaining a near constant total concentration of inorganic ions (ionic strength) and of specific ions. It must be able to control its volume and avoid excessive osmotic swelling or shrinking, especially before the advent of cell walls. It must have mechanisms to modulate its energy supply and the concentrations of the main internal organic chemicals – termed 'metabolites'.

This tightly integrated regulation is well illustrated by the process of turning the information in the DNA triplet code into a specified protein, which I alluded to earlier. This occurs via two major steps termed 'transcription' and 'translation' (see note 3). The blueprint for life exists in viruses, spores and even naked strands of DNA but *in vivo* the genetic information can only be translated accurately under specified conditions within a viable cell (see, for example, [16], but see also [17][18]). As I have emphasised, homeostatic mechanisms are required to maintain these conditions. The situation in a cell can perhaps be best visualised, as roughly analogous to an apparently stable eddy or whirlpool in a fast flowing stream, i.e. a structure created by an energy flow. More dramatically, tropical cyclones, be they labelled hurricanes or typhoons, are dynamic material structures powered by vast fluxes of energy drawn from sea water at temperatures exceeding about 26°C. Unlike homeostatically stabilised cells, the cyclonic structures are highly unstable and dissipate rapidly when passing over cooler seas or land that cannot provide sufficient heat energy inputs. Work is being done to generate the storm structure and the energy embodied therein is capable of hugely destructive 'work', as the inhabitants of many flattened Caribbean or Filipino communities know only too well. These are naive versions of much more sophisticated dissipative structures, as envisaged by Prigogine (see [19]). Just as in a persistent eddy or cyclone, a cell's internal environment appears constant but is, in reality, turning over constantly. This continuous replenishment is itself dependent on a flow of energy. But in biological systems, over the billennia of evolutionary time, mechanisms have arisen to store chemical energy, so helping to sustain the living entity in times of external turbulence. In Antonio Damasio's recent formulation [20], these mechanisms are geared to do more than just achieve continuity but have evolved to seek to ensure cellular and later organismal 'flourishing'.

As living systems have elaborated and competed with each other for resources, then the concept of homeostatic responses has to be extended to include an ability to withstand assaults by other organisms. Later in evolution they also included ensuring a coherent, internal balance between activities of different cell types or organs with a complex organism. I contend that comparable issues affect the interactions of social organisms including humans. All these homeostatic mechanisms, by definition, must be highly dynamic as well as creating a long-term, internal coherence. Indeed the term 'homeo-dynamic' might be more appropriate, cf. Damasio [20] and Rose [21], but in this essay I will continue to use the more familiar, albeit more passive, terminology.

Few scientists doubt that we are now in the middle of a new and fraught energy revolution. We are engaged in the painful process of re-engineering our economic and social lives away from a dependence on fossil hydrocarbon fuels (certainly their gaseous emissions) because of the growing threats from global warming and climate change [4][22][23][24]. This is occurring despite these fuels being the source of the energy that has driven, and indeed still drives the highly productive global agricultural, industrial and commercial revolutions. They have generated, for a significant proportion, but of course by no means all, of humanity, previously undreamed of material prosperity as well as a world of daunting physical and social complexity. They have, in turn, supported a dramatic rise in the human population. Consequently, our current responses to this challenge are conflicted by very powerful behavioural, social and economic tensions. Furthermore, other rapid changes, such as our growing digital and technical mastery and the emergence of globalised, interdependent, economies and political structures, are challenging many of the cherished certainties of the last half century.

Regrettably, despite the massive improvements in the living standards of billions, poverty remains pervasive in large parts of the globe. Vast numbers still lead hand-to-mouth existences, reliant on uncertain, often diminishing, natural resources and enjoying a very limited access to the affluent world and with little prospect of doing so. For these people, the sixth energy revolution has barely impacted. Even so, our species, particularly the affluent minority, has appropriated a huge and growing proportion of this world's natural resources to itself. Many argue that we are already exceeding our planet's capacity to sustain our still growing population. Nevertheless, human ambition, primed by aspiration and necessity and by rapid global information transfer, is undimmed. New energy sources must now be developed to power and sustain human society; this despite the continuing presence of adequate reserves of the fossil hydrocarbon fuels that have driven the sixth revolution. However, I contend that the trends revealed in this analysis of the long-term natural history of energy and associated phenomena suggest that technical innovations per se are unlikely to overcome humanity's problems. Simply seeking new, cheap, abundant, low-carbon energy sources will not suffice. I suggest that we are being compelled to a more radical reassessment of the relationship between energy, work, power, material and social complexity and humanity's place on this planet.

In this volume I will not discuss the astrophysical mechanisms that have produced both the raw chemical constituents of this planet and are the source of the Sun's incoming electromagnetic radiation (but see Lenton and Watson [10]). The latter makes this planet habitable. More extreme nuclear fusion reactions in large stars and supernovae are the sources of all the chemicals in the Periodic Table, a number of which are essential to living cells. Einstein in his famous equation ($e = mc^2$) rewrote the classical concepts of the conservation of energy and matter and showed that the fundamental physical process is the combined conservation of energy and matter and the *exchange of energy and matter*. A central theme of this volume is the evolution and elaboration of mechanisms to couple the work, enabled by the energy released by Einsteinian fusion in the Sun, to the building of material complexity on this planet. Although some of the same words are employed, these are entirely different phenomena.

I realise that extrapolating from physics and biology into human anthropology and more recent social history will be contentious. Nevertheless, I believe that a coherent picture emerges and that the simple notion of energy-enabling work and producing power, all modulated by homeostasis, is widely applicable. My hypothesis is an exercise in 'consilience', to borrow from the title of Edward O. Wilson's book [25]. I am arguing that the concept of sequential energy revolutions and related changes in homeostatic regulation over some 4 billion years not only offers a coherent narrative, linking biology, anthropology, economic and social history and human psychology, but a valuable perspective on our current political, economic and environmental dilemmas.

Readers not conversant with cell biology and biophysics may well find some of the concepts and evidence outlined in chapters II, III and IV daunting. These are the chapters dealing with the three primarily biological step changes. I hope that I have succeeded in conveying the most critical facts and hypotheses, and have included a short summary at the end of each chapter that may be helpful.

CHAPTER II

The Mysterious Origins of Life

The origin of the life on our 4.6 billion-year old Earth remains one of the great, unresolved mysteries. Microfossils and isotopic signals in rocks some 3.8 billion years (bys) old suggest the presence of living organisms. Much stronger evidence of bacterial fossil masses, stromatolites, has been dated to 3.5 bys ago. Scientists are searching for evidence of cell-like microstructures in ancient rock strata or the tell-tale trace changes in the isotopic ratios of elements, such as carbon, that characterise cellular metabolism [26]. But arriving at an approximate timing for the presence of life on this planet is only a small step forward; a much greater challenge is to propose a credible, comprehensive, hopefully testable, theory of how such a revolutionary event could have occurred.

We are faced with a formidable challenge of analysing how, when and even possibly why life has arisen on Earth and to assess the likelihood of it occurring elsewhere in the Universe. Was it an unique, chance event on Earth? An accident made possible by the accumulation of random events over tens of millions of years [27] or was it the result of predictable, possibly inevitable, processes arising from some of the fundamental facets of the physics and chemistry of matter [28]? In essence, an emergent property, which in various guises will have occurred elsewhere in the Universe. Some, of course, find the sheer complexity of the problem reason to cite divine intervention. Undoubtedly, the awe, which must accompany any serious consideration of this question, is all the greater when one realises just how complex, sophisticated and intricate are the structures and activities found in even the simplest cell. Every cell is bounded by a semi-permeable membrane with specific characteristics that enable it to selectively import some chemicals and exclude or export others. This membrane contains and constrains the dynamic living entity [16][29]. Each cell depends on several hundred specific and co-ordinated biochemical and physical reactions, each operating in a way that allows the total entity to survive and potentially replicate itself.

The issue of the origin of life (see note 5) can usefully be broken down into several discrete components. The first is the need to account for the abiogenic or pre-life origin of the complex organic compounds and the inorganic elements that are essential to any living entity, and, secondly, how these could have evolved into information-carrying and/or catalytic molecules capable of self-replication. Thirdly, there is the problem of creating the conditions to maintain a distinct entity separate from the external medium and able to concentrate and to retain the crucial molecules. Expressed somewhat crudely, this requires the formation of a 'bag' to hold the chemical goods and a source of energy that allows such a bag to maintain itself, i.e. to do on a microscale what a cyclone or an eddy cannot do. It must, as I have mentioned, sustain its own complex and dynamic biophysical and biochemical internal environment, far out of thermodynamic equilibrium with the external medium. In the Maynard Smith/Szathmáry formulation [9], these steps are considered as three separate revolutions.

A number of important experiments, starting with Miller and Urey in the 1950s [30], have concluded that, under the conditions that are likely to have existed on early Earth, the abiotic chemical synthesis of many chemicals associated with cell biology (amino acids, sugars, lipids and other vital chemicals such as nucleotides) would have occurred. However it is less clear how they could have polymerised, by losing water, to form long macromolecular chains. The precise chemistry of the early atmosphere is still contended but is likely to have been rich in the gases, methane, ammonia and hydrogen. Critically, our planet's size and distance from the Sun favoured the retention of water. Thus, as postulated almost a hundred years ago by Oparin and Haldane [31], it is quite probable that the chemical raw materials of life could have been synthesised from simple chemicals in the primeval Archaean or Hadean seas under the influence of heat and incident radiation. They also showed that fatty lipids spontaneously form little bubbles in water (or 'coacervates' to use Oparin's terminology). Thus, structures resembling cells would have occurred naturally. Over the years, the evidence for the prebiotic presence of the critical chemical building blocks of biology has been grown and even extended beyond our planet, such as to some of the moons of Saturn or Jupiter or even an asteroid such as Ceres [32].

However, a primordial chemical soup, even contained in a coacervate lipid sack, is still a long way in terms of both complexity and sustainability from a metabolising, replicating cell. It is speculated that as an intermediate between the prebiotic world and the current biology based on the DNA – RNA – protein axis, there was an intermediate RNA world. One reason for this idea is that the RNA group of molecules in modern cells act as a bridge between the information encoded in the DNA of genes and the proteins catalysing cellular activity. Also, RNA can act as both a replicator and a catalyst so suggesting that it might have supported both crucial functions in a primitive world [33]. This fascinating speculation is very important to our evolving understanding of early biological chemistry. However, it is not critical to this present volume, focusing on the energising of an ordered, complex, living entity.

It is useful, in my judgement, to distinguish between the energy required to promote the chemical synthesis of the specific biologically important chemicals, including the DNA and RNA bases and their nucleotides, in a prebiotic world and the energy required to sustain, indefinitely, an intact but dynamic replicating cell. The latter implies maintaining a cell line, indefinitely, out of equilibrium with its environment. The former chemicals are a prerequisite for life and, when polymerised and structured into viruses, plasmids or synthetic genes, contain the requisite information for effecting life. But, as I have emphasised, this information can only be expressed within a fully vital system requiring the highly specific conditions found only in a viable energy-dependent metabolising cell. In this volume I will focus on the vital step of energising that cell.

The traditional hypotheses are hard put to explain how these chemicals can become sufficiently concentrated and incorporated into a relatively stable vesicle. Any credible theory must also explain the source of the energy flux required to maintain, homeostatically, a distinct internal milieu. The huge differences between the complex intracellular macromolecules and the relatively simple organic compounds outside are well known. Less commonly appreciated are the differences in inorganic chemistry. While the Archaean seas are considered, as are today's oceans, to have been dominated by sodium (Na^+) and chloride (Cl^-) ions, the cytoplasms of cells tend to exclude these ions. Instead, they selectively concentrate potassium (K^+) and phosphate (PO_4^{3-}). Calcium (Ca^{2+}) is important as a micronutrient in cells but its close chemical relative magnesium

(Mg^{2+}) is accumulated to much higher levels. Thus there are clear biological-selective pairings of inorganic ions; these pairings appear to apply to all cells [34].

Two very obvious questions arise. Why is such ion specificity important to living cells? And how could it have originated? The first of these questions is the more approachable. For example, the presence of a specific K^+ concentration in cells can be related to the processes of converting the coded genetic messages in DNA sequence into individual proteins [17][18]. A messenger (mRNA) carrying the correct genetic information sequence, derived from 'transcription', must be 'read' (translated) and turned into the correct amino acid sequence, which, after folding and processing, defines the structure and metabolic function of that protein (see, for example, [16]). As mentioned, this 'translation' process takes place on structures known as ribosomes in the cytoplasm. Common salt, sodium chloride, even at modest concentrations, damages this process, and so must be actively excluded from the cell's cytoplasm. Both the synthesis of proteins and the maintenance of the 'correct' ionic balance require the expenditure of energy.

Recently Nick Lane in his book *The Vital Question* [35], building on the work of Mike Russell and William Martin and others, has advanced a coherent hypothesis to explain how living organisms may have arisen, emphasising the crucial role of energy. The hypothesis has also been summarised expertly by William Martin [35]. The energy, it is suggested, was derived from natural geophysical sources found within well-documented and ancient physical structures. The appropriate conditions for the evolution of, first, a proto- and, secondly, the fully capable cell(s) could have been met in the micropores of alkaline hydrothermal vents. In such microporous structures, local thermal currents have been shown, by a process called thermophoresis (see note 6), to concentrate the crucial abiotic building block, such as sugars, lipids and nucleotide bases, manyfold. These microporous structures, whose modern equivalents can be observed in vents in our current oceans, would, as well as having thermal gradients, contain natural gradients of pH, i.e. of protons (H^+) (note 6). These gradients would have been formed between the highly alkaline (high pH) fluid being vented from the lithosphere and neutral to mildly alkaline pH (~pH 8) of the percolating waters from the surrounding oceans or more confined waters. Such a pH gradient (a high pH signifies a low concentration of H^+ or protons) is a potential source of energy analogous to the gradient of negatively charged electrons (e^-) which we exploit daily in electrical circuits. Since these vents seem to be long-lasting, such a gradient could have provided, in Lane's hypothesis, a stable source of energy for the evolution of protocells.

One remarkable observation lending credence to his hypothesis is that proton (H^+) gradients, comparable to those in the alkaline hydrothermal and other vents, have been found near-universally across bioenergetic membranes of all living cells. A gradient of pH across a 'membrane', possibly initially the inorganic wall of a micropore, but later across a hydrophobic lipid biomembrane as in coacervates, would have created a proton motive force (pmf). Such a force is analogous to a gradient of negatively charged electrons in a conventional electrical circuit (emf) but consists of positively charged protons; both would generate an electrical potential gradient and a current. This electrical force would, and indeed still is, then used to drive the production of adenosine triphosphate (ATP). The nano-machine responsible is termed a 'proton ATPase' or 'ATP synthase', i.e. it is the enzyme that catalyses the synthesis of ATP from less energised compounds – ATP having what are colloquially referred to as 'high energy bonds'. This compound is the universal energy currency of living organisms and its synthesis via an electrically driven enzyme pump is one of the most important concepts in cell biology. When in the 1960s Peter Mitchell first produced evidence for this process in modern cells [36], there was considerable resistance to his idea. Some eminent scientists saw cellular metabolism purely in terms of organic chemistry and had spent many years searching for hypothetical chemical 'coupling factors' to explain the formation of ATP. Mitchell's insight was an important breakthrough. He created a new paradigm and a critical step towards the realisation of the tight bonds between cellular physics and chemistry.

Strikingly, the H^+-ATP synthase enzyme is coded by one of those genes whose origin can be traced back, by statistical molecular genetics, to the LUCA, i.e. the putative original cell from which, it is believed, all organisms are descended (see note 4). The realisation of the ubiquity and the lineage of the mechanism of ATP synthesis was a revolution not far short of the elucidation of the double helix of DNA. The beauty and flexibility of this mechanism is that, after escaping the confines of the original inorganic pore, the oxidation of a vast range of compounds, from simple sugars to reduced inorganic iron, could be and is today coupled to the generation of this membrane electrical gradient, to a proton motive force (pmf) and the synthesis of ATP. This system for energising cells is not only universal but has proved amazingly resilient and adaptable.

The thermal vents are also a very ancient source of hydrogen gas (H_2) due to a process called serpentinisation. In this series of reactions, promoted by heat, minerals from the Earth's crust that are rich in magnesium and iron react with water and release H_2 gas and secondarily methane (CH_4). These gases are themselves potent sources of chemical energy.

Sophisticated genetic detective work has sought to identify the genetic make-up of the LUCA. Martin and his team [35] suggest that some 'original' 355 genes can be identified and that these code for organisms similar to present-day H_2-using anaerobic chemolithotrophs – called methanogens (see note 7). Such organisms gain their energy, via the pmf and H^+-ATP synthases discussed earlier, and can build their cellular blocks exclusively from gaseous inorganic sources such as H_2, CO_2 and N_2 and other inorganic chemicals.

These observations are, of course, not proof that these mechanisms were crucial to the origin of life but are undoubtedly consistent, in physical, genetic and geochemical terms, with the supposition. I harbour one concern that the ionic strength and composition of modern cells are not readily compatible with LUCA evolving in highly saline oceans. The cellular inorganic chemistry seems to imply evolution in the presence of brackish seas. However whether or not Lane and Martin's specific ideas prove to be a lasting explanation of the beginnings of life, one can be confident that the twinned capacity to exploit the energy released from the controlled oxidation of a wide range of reduced substrates by the flow of protons and electrons down a chemical chain is fundamental to life. (This is termed a 'redox chain'; see note 8 for a brief explanation.) It is a matter of wonder that these redox and pH gradients and H^+-ATP synthases are at the root of phenomena as outwardly dissimilar as human respiration, algal photosynthesis and the ability of microbes to live in sunless mines.

This, I suggest, was the primary energy revolution that has, over the millennia, permitted life and defined this planet not only biologically but also atmospherically, oceanically and geologically. As I have stated, this energy-transforming and harvesting mechanism occurs universally in the three great domains of life, the *bacteria* and the *archaea*, both prokaryotic, exclusively single-celled, domains (but see note 9) with unstructured cytoplasms and no nucleus, and the other, more recent domain, the *eukaryotes*. These latter cells are characterised by complex internal structures (organelles). Many eukaryotic organisms are multi-cellular and fundamental to higher plants and animals, as will be discussed in chapter IV.

In emphasising the critical role of energy in enabling living organisms to exist, I am in no way seeking to minimise the importance of genetic information. However, without a cell structure, the information potential within DNA or RNA per se remains unfulfilled. Under many but not all circumstances, DNA molecules are remarkably chemically stable and inert. This trait permits reliable forensic evidence to be derived from tissue and body-fluid samples, some many decades old. Genetic data from DNA extracted and sequenced from fossils and even the odd tooth illuminate the evolutionary pedigrees of humans and other organisms. It is highly significant, as I have noted with ATP synthase, that the ribosomal translational mechanisms have been highly conserved over billennia [37][38]. Perhaps a good way of approaching this is to realise that the messages in the book of life have evolved dramatically over several billion years but the method of reading the book has remained remarkably constant.

The tight coupling and interdependence of the flows of energy and information deserve to be emphasised. Both the H^+-ATP synthase enzyme and ribosomal protein synthesis are among the mechanisms traceable back to the LUCA. The latter requires the energy supplied by the former, while the genetic information to synthesise the former is coded for in the latter. Although the genetic messages in the DNA sequences have necessarily altered throughout evolution, the mechanisms for their reading have remained remarkably constant. If this were not so, then the human capacity to genetically engineer the expression of genes from bacteria or fungi into plants or humans and vice versa would in all likelihood not exist!

The evidence suggests that the Earth's early atmosphere lacked oxygen and was dominated by nitrogen (N_2) and carbon dioxide (CO_2). It contained methane (CH_4) and hydrogen (H_2), so favouring reduced compounds such as ferrous iron (Fe^{II}) and reduced sulphur compounds. However, the redox mechanism (see note 8) for harvesting biologically valuable energy as ATP, as outlined above, requires not only a donor of energised electrons (i.e. ones with a high negative potential) but also an electron acceptor that has a much lower, possibly a positive potential. (The scientific convention refers to the energised donor as having a high negative potential and acceptor/recipient as having a lower negative or a positive redox potential; see also note 8.) The supply of the latter could well have been limiting in this early period.

Fascinatingly, this first biological energy revolution is revealed, geologically, in the deposits of ferric iron (i.e. oxidised iron) in huge ancient rock formations of alternating layers of reddish ferric-iron stained and dark carbon-rich shales dating from ~3.5 bys ago (see [10]). These observations also imply that, originally, there was plenty of exploitable dissolved ferrous (Fe^{II}) iron around in the early oceans to energise and sustain life (see Figure 1).

From the perspective of this volume, the crucial point is that a continuous flux of free energy, derived from redox gradients, would have been essential for living organisms to emerge, be sustained and evolve. This flow allowed and still allows cell structures to be maintained out of equilibrium with their immediate environment. Schrodinger in his famous book *What is Life?* [39] wrote, somewhat misleadingly, of life creating negative entropy (order out of disorder (high entropy)) through an energy flux, but at the expense of an increase in the entropy of the external environment. As I noted earlier, this is much like a quasi-stable eddy occurring in a fast-flowing stream (see also [19]). Conceptually it may be easier to think simply in terms of ordered structures including DNA and proteins being created at the expense of a continuous flow of free energy. Such ordered structures then depend on a range of homeostatic mechanisms to sense and respond to changes in their external medium in order to retain their internal dynamic integrity. This is of course in contrast to an apparently stable eddy in a stream, which will dissipate rapidly if the flow rate in that stream changes.

A protocell would have 'died' without a constant flow of energy. Even today, without homeostatically induced stability and energy storage, this would be the fate of any cell. However, biological change has promoted the evolution of sophisticated homeostatic mechanisms, and, in its 'inventiveness', found many ways to store chemical energy and/or put cells into 'hibernation' so as not to die while awaiting more favourable conditions for life to flourish.

Early prokaryotic life would have been dependent on and, possibly after many hundreds of millions of years, been constrained by the availability of dissolved reduced organic or inorganic compounds, e.g. Fe^{II}, or gases, e.g. H_2 or methane (CH_4), as energy sources. However, since the original atmosphere may have been mildly reducing and, certainly, oxygen-free, the availability of appropriate electron acceptors at the base of the redox chain might have been as problematical. In terms of their biochemical capability and agility, the prokaryotic bacteria and archaea were and, indeed, remain supreme [40]. Their success has been based partly on their ability to use a great variety of energy sources to generate their ATP and on an energy-intensive biochemical mechanism to fix atmospheric CO_2 so synthesising the cell's organic constituents. In parallel, some bacteria and archaea evolved a further energy-demanding mechanism to fix atmospheric nitrogen into ammonium (NH_4^+)and thence into organic forms [41].

The versatility of the prokaryotes has been greatly helped by their ability to transfer individual or packets of genes laterally from one organism to the next. This type of rapid DNA/information transfer may allow much greater evolutionary dexterity and flexibility than is permitted in classical Darwinian selection that depends on sexual reproduction and the survival of the fittest progeny to, in turn, reproduce the next generation. The dynamic of Darwinian selection is based on competitive reproductive success. But, given the ease of lateral gene transfer and the absence of sexual gene mixing in prokaryotes, the dynamic may be subtly different. Amazingly, these single-celled prokaryotes were the only living organisms on Earth for some 2 bys, almost half of its history (see Table 1 and Figure 1). To this day they remain major contributors to the biosphere in terms of the total number of individual organisms and the diversity of species (cf. [42] and [43]). Estimating the number of both pro- and eu-karyotic species, still more the total number of organisms living at a particular time, is fraught with problems and much contested. In daily life and television adverts, 'bacteria' are often seen as grotesque threats, as they carry diseases and types of food poisoning (e.g. salmonella or *E. coli* poisoning). Nevertheless, they are essential, often in unexpected ways. Remarkably, we carry in our healthy bodies about as many 'foreign' prokaryotic cells as our own 'native' cells; indeed, our health is dependent on a symbiotic relationship between our human genome and the much larger, foreign prokaryotic genome mainly in our guts.

To summarise, the energising of prokaryotic cells depends on the formation of ATP molecules in a process mediated by a proton and electrical gradient derived from the controlled oxidation of reduced chemicals. This mechanism may well reflect the original source of energy for all living cells in the natural proton (pH) gradients in oceanic or brackish alkaline vents. Overall, energy sources have been exploited to sustain self-replicating living cells of amazing internal biochemical complexity, but whose material structural complexity, i.e. that are easily visible by light or even electron microscopy, is limited. They are almost exclusively single-cell organisms (see note 10). Even over the four billennia none have evolved into true multi-cellular organisms. All have refined an array of homeostatic mechanisms to sustain their intracellular integrity. By initiating and enabling such cells, this energetic innovation has not only set life on a pathway leading ultimately to the great diversity and wealth of biology, including *Homo sapiens*, but has also shaped much of geology.

After a lapse of at least another billion years, a second great energy revolution occurred.

CHAPTER III

Harvesting the Sun

Many ancient societies such as Pharaonic Egypt, the Aztecs and the Celts have worshiped the Sun. All agricultural societies have been and remain profoundly aware of their dependence on the Sun.

In the words of the Reverend Eli Jenkins:

> Oh let us see another day!
> Bless us this night, I pray,
> And to the sun we all will bow
> And say, goodbye – but just for now!
>
> (Dylan Thomas)

The daily cycle of light and dark and the seasonal reawakening in the spring sunshine and dismay at winter's darkness and decay have always ruled existence. Even today in our highly urbanised and electrified society, in the absence of sunshine we may suffer from SAD (seasonal affective disorder) and, in some more extreme instances, vitamin D deficiency and rickets. Our dependence on the sun's radiant energy remains fundamental, although masked and almost forgotten in our urbanised world.

It is suggested that even as early as 3.4 bys some bacteria, termed anaerobic phototrophs (see note 7), were able to make use of solar energy to promote specific biochemical processes. These likely involved the oxidisation of reduced forms of sulphur or iron and other similar inorganic and organic compounds. The reactions were coupled to a compound capable of accepting the electron at lower energy level (the so-called terminal electron receptor) producing a final product such as elemental sulphur or reduced iron (Fe^{III}) (see note 8). The advent of larger quantities of Fe^{III} would have increased the availability of electron acceptors for expanding non-photosynthetic life. Organisms using such chemical sources of energy are called chemotrophs (see note 7). These reactions can create the proton motive force (pmf) and electrical gradient across the cell's membranes that, as was discussed in the previous chapter, can be used to generate ATP. Nevertheless, in all probability, the sources of both the donors of energised 'reduced' electron donors and more 'oxidised 'acceptors of those electrons would have been limited, thus living organisms were, in effect, in an energetic cul-de-sac, unless a new source of energy was accessed.

Fast-forwarding a billion years, by 2.7 to 2.5 bys, evidence emerges of the presence of the cyanobacteria (cyano = blue-green). These bacteria evolved and still maintain to this day a full photosynthetic capacity [44]. That is, they were, and are, able to use the energy in the photons (light) from the sun to excite/energise electrons released from green pigments, such as chlorophyll, and to catalyse the splitting of water into its constituent parts – that is, into protons (H^+), electrons (e^-) and oxygen (O_2), with the latter released into the atmosphere [45][46][47] (see Figure 2). After their excitation, the electrons and protons are fed into essentially the same redox mechanism for energy capture and ATP synthesis that was discussed earlier. The chemical energy so produced is then used to capture atmospheric carbon dioxide (CO_2), using a mechanism whose evolution long predated photosynthesis. A biochemical mechanism also evolved in some cyanobacteria to fix atmospheric nitrogen into organic compounds. Together they comprise a set of reactions that can generate the all-organic building blocks of cells from very simple and common chemicals on the planet's surface by exploiting incident solar energy. The acquisition of the light-harvesting systems, usually referred to as photosystems I and II, itself involved a complex set of events, possibly requiring the fusion of light-trapping pathways derived from two types of prokaryotic

Figure 2: Earth's oxygen timeline – the breath of life

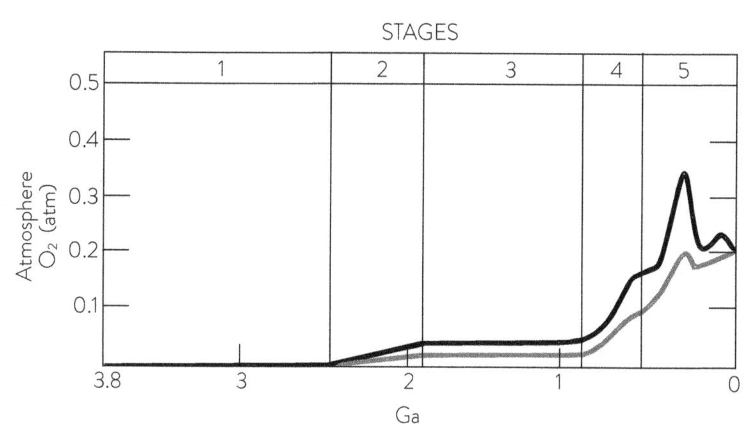

Two estimations of oxygen (O_2) build-up in the Earth's atmosphere over some 4 billion years [45].

The upper and lower lines represent the range of the estimates, while time is measured in billions of years ago (Ga). (Note 'giga' and 'billion' have identical meanings = $10\times^9$)
Stage 1 (3.85–2.45 Ga): practically no O_2 in the atmosphere.
Stage 2 (2.45–1.85 Ga): O_2 produced, but absorbed in oceans and seabed rock.
Stage 3 (1.85–0.85 Ga): O_2 starts to gas out of the oceans, but is absorbed by land surfaces and formation of ozone layer.
Stages 4 and 5 (0.85 Ga–present): O_2 sinks filled, the gas accumulates.

organisms and is documented in Lenton and Watson [10]. It did not exploit directly the light-harvesting mechanisms of chemotrophs.

This solar-energised splitting of water (H_2O) has literally revolutionised our planet, allowing life forms to multiply and diversify. It has re-engineered our planet by changing its atmosphere from being neutral or slightly reducing to one in which oxygen slowly became a major component (see Figure 2) [45]. This in turn gradually oxygenated much of the surface of the world's oceans and caused major changes in the surface geology by promoting the formation of new types of sedimentary rocks. However, in overall energetic terms this meant that living cells had succeeded in tapping into a near limitless source of energy, one which has remained the dominant driver of life on Earth ever since [46]. Life on this planet was released from any general dependency on pre-existing or abiotically produced energy-rich sources.

One paradoxical consequence of the time-sequence of these evolutionary events revolves around the enzyme (called Rubisco (Ribulose-1,5-bisphosphate carboxylase-oxygenase)). This enzyme, near universally, brings about the net fixation of atmospheric CO_2 to yield sugars [47]. However, it not only reacts with atmospheric CO_2 (i.e. the positive function on which all life depends), but also catalyses a reaction between a sugar and O_2. The latter function leads to series of reactions known as photorespiration, during which there is a net loss of CO_2 to the atmosphere, i.e. a sort of respiration. The relative importance of these two reactions depends on the comparative concentrations of CO_2 and O_2 at the active site of the enzyme.

This situation has had paradoxical consequences (see also note 11) and for those who believe in a Creator God suggests s/he had a wry sense of humour or was not all that prescient, failing to anticipate that an oxygenated atmosphere would 'interfere' with

Rubisco. As I will discuss much later, this relationship has become something of a cause célèbre in the minds of climate change deniers. They argue that 'fertilising' the world with more and more CO_2 can only be beneficial and can help 'cure' the error of oxygenase 'inefficiency'. Despite the naivety of this attitude, the catalytical duality of Rubisco is a major challenge to plant science as it contributes to the loss of yield of some of mankind's major crops such as wheat, the source of the daily bread of a large slice of humanity, under environmental stress.

As I have noted, the first great oxygenation event started to increase the atmospheric O_2 levels significantly but modestly (see Figure 2). They reached only about a quarter of current levels after half a billion years. Over this period, gradually and incrementally the Earth's geochemistry and biochemistry were transformed. Oxygenating the seas and atmosphere led to the formation of many new minerals and a decline in the levels of freely available reduced iron and other similar compounds, as these can be oxidised chemically as well as biochemically. This, in turn, reinforced the importance of the capture of light energy and the role of cyanobacterial photosynthesis.

While new sedimentary deposits can be detected geologically, an equally important biochemical revolution was precipitated within cells, one that can now be detected by evolutionary genetics (see also Figure 1). The advent of atmospheric O_2 made more energy efficient, fully aerobic respiration possible. That is the controlled oxidation of the reduced organic compounds, e.g. sugars, again coupled to the pmf based-mechanism outlined earlier, using the reduction of O_2 to H_2O as the terminal electron acceptor (note 8). While growing on sugar and other organic compounds in the absence of oxygen, some organisms can, and indeed still do, use various terminal electron acceptors in a process known as 'anaerobic respiration'. These acceptors include using the nitrate/nitrite couple (NO_3^-/NO_2^-). But since nitrite (NO_2^-) is a potential toxin and a cause of blue baby syndrome, it can be problematical. The energy yield from these reactions is relatively low and the dispersal of the final product, e.g. nitrite or sulphur, a potential constraint on growth. So-called nitrate respiration yields about half as much energy, measured as ATP production, compared to full oxidation coupled to reduction of O_2 to H_2O (see also note 8). Other electron acceptor couples are even less efficient. When the O_2/H_2O couple is used, not only is the energy yield higher but the water released disperses readily. An even more dramatic energetic comparison is with organisms growing by anaerobic fermentation to yield ethanol or lactate. This process, while relatively recent in evolutionary terms, is of abiding interest to humans but is highly energy inefficient, yielding only about 2 moles of ATP from a mole of glucose, whereas aerobic metabolism coupled to O_2/H_2O produces 34–6 moles of ATP: a possible 14-fold gain in energy efficiency. We are nevertheless partial to some of the products of fermentation.

The photosynthetic splitting of water (photolysis), so releasing chemical energy and oxygen gas and allowing the fixation of CO_2 to sugar is therefore neatly mirrored by aerobic respiration. In the latter, the oxidation of sugars lead to the uptake of oxygen and its conversion to H_2O with the generation of energy as ATP. The CO_2 gas, produced by the oxidation of these reduced organic compounds, is released back into the atmosphere,

The combination of true photosynthesis and global oxygenation was the basis of the second revolution and led over millennia to the gradual emergence of a wide variety of prokaryotic organisms and short prokaryotic food chains. Much later, as will be discussed, complex eukaryotic cells and multi-cellular organisms emerged, the large proportion of which feed off the primary photosynthesising organisms but were, and remain, dependent on this highly efficient aerobic respiration for their energy.

In all likelihood, these historic biochemical and atmosphere events precipitated a series of great global crises. Firstly, the oxidation of atmospheric methane (CH_4), a very potent greenhouse gas (GHG), to CO_2, a less effective but still important GHG, is thought to have triggered what is called the first 'snowball earth' event. This event in which this planet was cocooned in ice may well have lasted hundreds of millions of years (see Figure 1).

Secondly, despite its energetic advantages, gaseous O_2 is highly reactive and potentially damaging to cells because its presence readily leads to the formation of toxic, free radicals. Consequently, organisms have had to develop defence mechanisms to protect cells from oxygen gas in the atmosphere – mechanisms we see today in modern organisms and are also the background to consumer interest in the anti-oxidant content of our foods. As the air and seas gradually became aerated, the reduced forms of inorganic iron and sulphur that had been, like methane, important as major energy sources in the earliest years would have been oxidised, thus limiting their availability. The composition of sedimentary rock deposits also changed in turn. Some billion years later, a second great oxygenation event took place whose immediate cause is unclear. Over millions of years the atmospheric O_2 levels (after some further ups and downs) also rose to the current level of circa 21% (see Figure 2).

In summary, in the second revolution a cellular mechanism evolved to capture the energy of sunlight and to use it to capture atmospheric CO_2 and convert it into organic compounds while also emitting oxygen gas into the atmosphere. This, in turn, allowed energy-efficient respiration to evolve. The latter process exploits the chemical energy in sugars to enable cell growth and replication while releasing CO_2 back into the atmosphere and converting atmospheric O_2 back to water. In this revolution the observable material changes in the structure of cells were small and mainly related to the modification of the membrane structures already emerging in pre-photosynthetic cells. No doubt additional homeostatic mechanisms would have been required to stabilise the new life forms – most obviously to allow organisms to seek out light. Less obviously, regulatory mechanisms would have been needed to prevent too much light and energy damaging the cells. Nevertheless, an energy source was captured that both rid life of its dependence on limited resources of electron-donors and receptors and paved the way to a new global stability and diversity. The great global carbon, nitrogen and oxygen cycles were initiated all dependent on reliable and abundant solar irradiance. But after a further billion years, there came a third revolution.

CHAPTER IV

A Structural Revolution – Complex Cells

The role of energy in first enabling and then sustaining cellular life on this planet is not in doubt. Prokaryotic life flourished for billennia and no-one questions the fundamental importance of photosynthesis in energising life and framing our planet's atmosphere, hydrosphere and lithosphere. In contrast, the critical role of energy in the third revolution has only recently become apparent and is, in many ways, subtler, even contentious.

For some 2 bys only morphologically very simple, microscopic, bacterial and archaean cells can be detected in the palaeontological record (see Figure 3(a)). By about 2 to 1.7 bys ago, traces of cells with a much more complex internal structures are observed in thin sections of some rocks [48]. These cells have been given the name 'eukaryotes' – meaning cells with a true nucleus. It is clear that all higher multi-cellular organisms from animals to vascular plants and from macro-algae to fungi are composed, exclusively, of such cells (see, for example, Figure 3(b)). Many single-cell organisms – protists – such as amoeba and yeasts, are also eukaryotes but *all* complex organisms are eukaryotes.

As well as possessing nuclei, such cells exhibit an internal structural complexity quite unlike that found in prokaryotes. The nucleus itself, containing genes folded into distinct chromosomes, is bounded by a double membrane. Other important internal membrane-bound structures, known as organelles, include mitochondria, endoplasmic reticulum, Golgi bodies and vacuoles (see, for example, [35]); later, chloroplasts were added to this list [16] (see also pages 39 and 40). Each organelle forms a distinct compartment within the cell. Such eukaryotic cells, which are usually at least 10 times larger than their simpler, smaller prokaryotic brethren, also have a dynamic, mobile cytoplasm. They have straight chromosomes in their nucleus and the capacity for meiosis and mitosis (see note 12) and for true sex.

Amazingly, the genetic evidence traces the ancestry of all current complex eukaryotic organisms to one single ancestor. To date, no intermediary stages or forms have ever been found, i.e. we are aware of no part eukaryotic/part prokaryotic cells. Despite the immense profusion and diversity of life forms and of biological habitats, from oak trees to lions, from insects to fish, from coral reefs and savannah to acid moorland, eukaryotic cells and their consequentic complex life forms appear to have evolved once only. So, there can be no doubt that this was a singular event of colossal global significance. As eukaryotes ourselves, in relation our perceptions of humanity's place on this Earth, arguably this was as significant an event as the much-debated origin of life, although it has not been part of the theological disputation! The biological record implies that, without this event, complex and certainly sentient life could not have emerged.

A broad consensus has emerged that this massive evolutionary, *singular* step involved the one-off absorption and retention of a donor bacterial cell by an archaean cell. The former evolved to

Figure 3: Internal structural complexity of (a) prokaryotic and (b) eukaryotic cells

(a) Prokaryotic cell

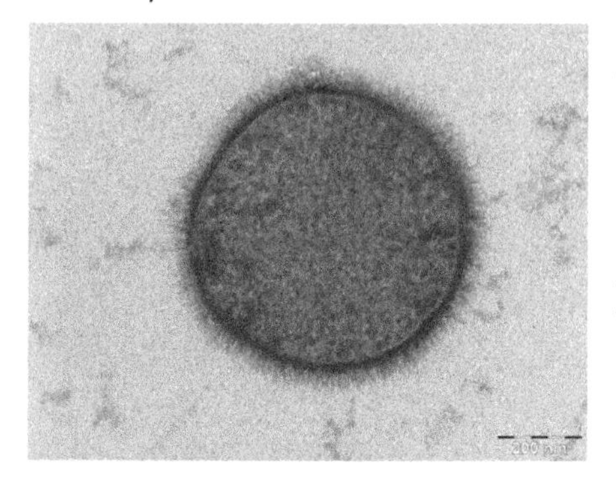

Micrograph of a cell of *Bacillus subtilis* about 600 nm (10^{-9}m) in diameter. The cytoplasm shows internal 'granulation' due to ribosomes and naked DNA strands but no discrete genet. The cell is bounded by a cell membrane and cell wall with external mini flagella to allow movement.

(b) Eukaryotic cell

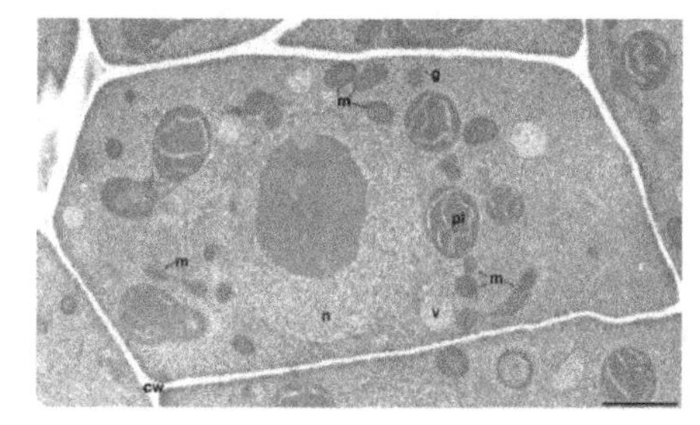

Young cell from plant *Arabidopsis thaliana*. The cell is bounded by a cell membrane and cell wall (labelled 'cw') The nucleus 'n' is bounded by its own double membrane, a number of mitochondria are visible in the cytoplasm as are Golgi bodies 'g' and plastids 'pl' that mature into chloroplasts. Small vacuoles 'v' are visible, which will develop, as the cell matures, expands and extends, to occupy some 90% of the volume of the mature cell that will be as much as μm (10^{-6}m) long.

became the mitochondrion and the receptor archaean cell contributed much of the rest of the cell's architecture [35][49]. Such a process is termed 'endosymbiosis'. A comparable event did occur several hundred million years later when a prokaryotic cyanobacterium was ingested into an existing line of eukaryotic protist cells to create a chloroplast. This organelle transmitted a photosynthetic capacity into eukaryotes [44]. This second endosymbiotic event was itself highly significant, leading to lower and higher plants such as algae and flowering plants (angiosperms) and ultimately to the colonisation of land. Even so, its significance cannot be compared with the primary evolution of eukaryotes themselves.

While it is generally agreed that the great revolutionary evolutionary step was the endosymbiotic incorporation of a bacteria into a recipient archaean cell, the mechanisms by which it could have occurred and its implications are a focus of current research and speculation. Mitochondria are often referred to as the 'power plants' of eukaryotic cells. They are able to synthesise ATP at rates sufficient to energise the whole cell while containing only a small number of their original bacterial genes.

Nick Lane [35] suggests that the fundamental characteristic of the eukaryotic revolution is revealed by thinking in terms of the energy available per gene in a given cell. In bacteria some 75% to 80% of a cell's energy budget is devoted to ribosomal protein synthesis. This process that, as well as being highly conserved as previously noted, is central to both the cell's general metabolism and reproduction and to its regulatory homeostatic mechanisms. When expressed in the conventional terms of 'metabolic rate', a gram of prokaryotic bacterial cells may consume oxygen [35] at a rate three times faster than of a gram of their eukaryotic cousins. But according to Lane, this is highly misleading. The cell and genome sizes in pro- and eu-karyotic cells differ so drastically that, according to Lane's calculations, an eukaryotic cell may have some 200,000 times as much energy available *per gene* as a prokaryote. This, Lane argues, has been achieved not only by an archaean cell ingesting a bacterium to create a mitochondrion, but crucially by the latter shedding virtually all the original bacteria genome. The 'endosymbiont' rising from the captured bacteria retained only a small cluster of 13 ex-bacterial genes. Significantly, these are essential for the close in situ regulation of the redox electron transfer chain and proton ATP synthase found in the mitochondrion. The proto-eukaryotic cell had, de facto, acquired a small internal 'power plant' without having to use the ATP output of that power plant to fund all the normal protein synthesis associated with the incorporated 'cell'. Eukaryotic cells have multiplied this benefit by enlarging their 'power-load' and by carrying multiple gene-cheap, energy-enriching mitochondria. A startling example of the multiplication of mitochondria is found in highly metabolically active cells of the human liver that contain as many as 2000 mitochondria each. In cells with a very large energy demand, such as cardiac muscle cells, these organelles may occupy up to 25% of the cell volume. It cannot be overemphasised that the symbiotic fusion of a bacterial and archaean cell to create an eukaryotic cell remains a singular event: only observed once in billions of years of the Earth's history. It follows that, in all likelihood, it was both a highly improbable event and one depending on an unusual, possibly unique set of biological circumstances.

It must be emphasised that prokaryotes have continued to flourish to this day, displaying their impressive biochemical dexterity, but that the eukaryote revolution primed the world for new possibilities and new complexities. This revolutionary step prompted Nick Lane to write [35]:

> Since the first eukaryotic cells arose some 1.5 to 2 bys ago, we have had warfare, terror, murder and bloodshed: nature red in tooth and claw. But in the preceding aeons we had 2 billion years of peace, symbiosis, bacterial love and what did this infinity of prokaryotic life come up with? (p. 157)

In his view – very little! This revealing passage of purple prose emphasises not only the significance of the biological step change but implies that, despite their biochemical precocity and adaptability, to the human eye prokaryotes compare unfavourably with the structural sophistication, physiological complexity and diversity and beauty of eukaryotic life! So, what did the latter come up with?

Clearly the eukaryotic revolution allowed a much greater investment in complex structures. Over time it gave rise to sex, to multi-cellularity, to cell differentiation and specialisation and to the allocation and integration of physiological tasks in specialised organs. Indeed, as the quotation suggests, it opened the door to the huge new scope of neo-Darwinian selection based on sexual gene exchange and reproductive competition and to increasing ecological complexity [35][50][51].

It is not clear how long the process of bacterial/mitochondrial assimilation and cellular adaptation took or whether there were a number of false starts before one succeeded. It appears that over the next billion years these changes were consolidated but with little apparent change in the fossil record (see Figures 1 and 2 for timelines). Nevertheless, this long hiatus, sometime dubbed 'the boring billion', must have supported crucial changes in cellular metabolism and cytology. Some 600 to 700 million years ago there is evidence of a series of 'snowball earth' events, followed by a second, greater, atmospheric and oceanic oxygenation event. About 540 million years (mys) ago, startling changes in fossil record have been observed called the 'great Cambrian evolutionary explosion' [52]. At this period there emerge the paleo-skeletons of some very strange multi-cellular organisms as well as the basic structures of many of the complex organisms we find today (cf. [50][51][53]). One of the most dramatic examples of this 'explosive' event has been observed in Burgess Shale of western Canada and is described vividly by Stephen Jay Gould [51].

Some speculate that the great Cambrian explosion was stimulated by the rapid rise in the level of atmospheric oxygen. This rise might well have increased the energy yield from respiration but, as far as I am aware, the hypothesis is unproved. Nevertheless, a highly significant step in the evolution of animals and later plants and the origin of much of their diversity occurred at this time. Whether this event should be considered a separate energy revolution appears unclear and, in the absence of much more information, I have not chosen to treat it as such. It was also clear that this was not a 'singular' event like its predecessors; in all likelihood there were several roads to complex multi-cellularity. The final resolution of this conundrum must be left to future research.

Another major evolutionary event came with the colonisation of land by plants (~450 mys). Again this step had important energetic implications. In all probability, the first colonisers of land were lichens – a symbiotic cohabitation of prokaryotic blue-green 'algae' (cyanobacteria) or true algae with eukaryotic fungi. Today, such remarkably robust symbionts and cyanobacteria themselves are usually the first colonisers of bare rock and appear to exist on fresh air – indeed, they obtain many of their nutrients from the atmosphere through rain and their energy from sunlight. The colonisation of land by plants per se may have involved a number of steps including a symbiotic relationship between the plant and specific fungi that help the former access the nutrients from the soil substrate. Over time, the successful colonisation of land opened the way to a significant increase in the global photosynthetic capacity as well as providing other non-photosynthetic creatures with new habitats.

Gradually over several hundred million years terrestrial and oceanic ecology became more recognisably related to, but, of course, still very different to, that found today. It is worth emphasising that, despite the critical step of the colonisation of land by plants and animals, some 50% of global photosynthesis, with its contribution to the turnover of atmospheric and terrestrial carbon and oxygen, as well as energy capture, is carried out by small marine algae and cyanobacteria [53]. Similarly, much of planetary respiration is undertaken by prokaryotes, not, as we tend to assume, by animals and higher plants. The great global geochemical nitrogen, carbon, oxygen and sulphur cycles underpin both terrestrial and marine life and allow the planet to remain out of thermodynamic equilibrium with its surroundings. These cycles depend heavily on prokaryotic organisms with more modest contributions from eukaryotes.

Throughout these aeons many layers of sedimentary rocks were laid down in addition to the extrusion of igneous material from the mantle. The chemistry of the new sedimentary rocks, shales, sandstones, limestones and chalks was influenced and, in some cases, determined by the biology. After all, the raw materials for sedimentary chalk and limestone are the shells of small, dead marine creatures. These and some other rocks have acted as sumps for a small proportion of the carbon captured by the photosynthesis being carried out by the flourishing lower and higher plants and cyanobacteria. Important to this volume, this led to residual sedimentary hydrocarbons reserves being accumulated over hundreds of millions of years. These included the carboniferous coal measures of south and north-east Wales, and the global oil and gas reserves.

The emergent ecological chains depended not only on the conversion of solar energy into matter but the further conversion of the chemical energy embodied in this matter – in carbohydrates, fats and proteins – to empower each subsequent link in the food chain. Simplistically, the energy travels from primary carbon fixer to fodder to herbivore to carnivore or omnivore, while each death also provides the substrates for abundant microbiological life. And as poetically envisioned by Philip Pullman in his trilogy, *His Dark Materials* [54], death sustains life and the cosmic release of atoms sustains our Universe, just as our life depends on chemical elements from beyond our solar system.

Inevitably, ecological food chains are energy-flow chains [7]. At each step, the energy availability declines steeply with only about a tenth of energy being available to the next link in the chain. In general, the higher the initial energy input (providing that water and other nutrients are not limiting), the greater the biological diversity that the habitat can maintain. Some non-photosynthetic food chains still exist, much as in the world after the first revolution some 4 bys ago. They are found, for example, in sunless mines and caves, but in terms of quantitative global energy fluxes they are trivial compared with photosynthetically-derived fluxes. Nonetheless, they are of great practical and academic scientific interest. For a local perspective, it is worth noting that such short chemotrophic energy chains have been found in mines in Snowdonia as well as in more exotic locations and are based on exploiting the chemical energy potential of reduced iron (Fe^{II}).

The emergence of the eukaryotic domain created new regulatory requirements. When discussing the primary energising of life, the fundamental requirement for homeostasis to maintain a specific, yet dynamic internal environment – a healthy cytoplasm – was emphasised. It was also stressed that all cells need to have the capacity to respond to external opportunities and threats. However, even in the simplest eukaryotic protist, there are not one but many membrane-delineated, internal compartments. The content of each had to be kept metabolically and dynamically within specific bounds. Each has its role and needs, which must be co-ordinated with all the others and integrated into the homeostatic regulatory responses of the whole, albeit single-celled, organism.

I will return to this issue latter but suffice to illustrate the issue with two contrasting examples. As might be anticipated, very precise mechanisms exist to export ATP from mitochondria and chloroplasts for use in the whole cell. The ATP is generated from adenosine diphosphate and inorganic phosphate by a H^+-ATP synthase located on the inner membrane of the organelle, exploiting a proton gradient, as previously noted. Another enzyme, called a 'translocase', then exchanges 'spent' ADP from the cytoplasm for the newly formed, energy-rich, mitochondrial ATP to keep the cell energised and functioning. This process is tightly regulated but works amazingly quickly as cells can use about a million molecules of ATP per second to maintain their functionality [35]. A problem worthy of modern control engineering!

Secondly, some mature plant and algal cells, be they single-celled (e.g. *Valonia ventricosa*) or in multi-cellular angiosperms, are dominated by very large vacuoles. These organelles are essentially a large liquid-containing sack, one of whose functions is to allow the organism to achieve a larger volume and greater surface area to intercept light, water and nutrients without too heavy a material and energetic investment in the complex biochemicals and exacting ionic demands of the cytoplasm [2]. There is now a wealth of data on how the homeostatic relationships between the vacuole and the whole cell are regulated and modified in response to internal and external changes such as drought and salinity.

Adapting to the presence of internal organelles was but a step upon the long regulatory homeostatic ladder. As mentioned, multi-cellular organisms are all eukaryotic. (A minor exception/anomaly is discussed in note 10.) They must, however, have developed homeostatic mechanisms to regulate and co-ordinate the relationships between the cells that comprise the integrated whole organism.

Over time, the sophistication of these mechanisms must have increased as the various cells became specialised and specific organs, e.g. liver or petal, developed with their own functions. As would be anticipated, the specific problems posed, and regulatory methods used, are different in animals and plants.

However, the critical point is simple: cell co-operation, the division of labour and functionality, and the mind-boggling range of biological forms arising from that co-operation, would not have been possible without a parallel evolution of homeostatic control mechanisms. These secured the efficient functioning and co-ordination and wellbeing at the organelle, whole cell, organ and whole organism levels and have led to structures of exquisite beauty and complexity.

The mechanisms for regulation and co-ordination are themselves remarkably sophisticated. Without such mechanisms such as electrical, especially membrane-based, nerve impulses in, but not exclusive to, animals and chemical/hormonal signals (in plants in the phloem or xylem and hormonal transport in animal fluids) and physical links (e.g. plasmodesmata in higher plants and algae), no complex higher organisms could exist. However, the primary selection process leading to these sophisticated integrative systems and this immense biodiversity has been competition. Natural sexual selection and the survival of the fittest is itself a dominant eukaryotic trait. In biology there are many examples of extreme cruelty and events characterised so dramatically in the earlier quotation from Lane and in phrases such as 'the selfish gene' [9] and 'nature red in tooth and claw'. However, it is as well to recall the symbiotic nature of the formative eukaryotic event and that cellular co-operation underlies multicellularity. I will return to these issues in a later chapter.

The prokaryotic and eukaryotic domains have continued in their parallel but partially interdependent worlds for many hundreds of millions of years. Naturally, the prokaryotic bacterial and archaean domains have taken full advantage of new habitats offered by the diversity generated by eukaryotic evolution. The neo-Darwinian selection pressures in the eukaryotic domain have been influenced by many external events. These include the vast tectonic movements of continents, the Milankovitch cycles in the orbit of the Earth around the Sun leading to a series of ice ages, comet collisions and the other geological and climatic upheavals.

The gradual evolutionary changes were punctuated by a series of mass extinctions. Five major extinctions are commonly recognised, with that at the K-T or K-Pag (Cretaceous/Tertiary or Cretaceous-Palaeogene boundary), which killed off the dinosaurs (some 66 mys ago), being the best known. In the late Permian era (252 mys) an enormous extinction event obliterated perhaps 95% of all organisms. We are the descendants of the lucky few that survived these cataclysms. It seems a reasonable assumption that this cycle of life and death, ascendency and decay would have continued to this day without the impact of another major energy revolution.

To summarise: with the evolution of energy-enriched eukaryotic life, a step change can be observed in the material complexity of cells and their biological potential. The stabilisation and full expression of this potential appears to have taken about a billion years to emerge in the Cambrian explosion. This heralded the emergence of a wealth and diversity of complex, multi-cellular life forms. These then have evolved under the competitive pressure over a few hundred million years into the startling biodiversity and range of habitats that characterised this Earth. It is the 'recapitulation' of this evolution in the DNA sequences within organisms that has provided such exciting evidence of the complexities of evolution [12][14][50] and of the singularity of the formation of a first viable eukaryotic cell, the original eukaryotic common ancestor.

These huge evolutionary changes in biological forms and structures required a parallel, but perhaps less obvious, evolution in the homeostatic mechanisms required to ensure the stability and integrity of the structures and co-ordination of activity and energy flow from the level of subcellular compartments (organelles) through to the whole organism and, as we will discuss later, into communities of organisms.

The interdependence of geology and biology has remained tight. The significance of pre-eukaryotic life has remained undiminished in, for example, the global nutrient cycles. Neither the importance nor the diversity of prokaryotic life has diminished. Rather, prokaryotes found new habitats to colonise, including within large eukaryotic organisms and new forms of symbiosis have also emerged. This period also saw the beginning of work being directed to the creation of social as well as material biological complexity, as illustrated by the social insects.

CHAPTER V

The Hominid Factor

Only some 2 million years ago on Earth's timeline (see Figure 1), a fourth great energy revolution was set in train. Initially it appeared to be of limited scope, even at first glance trivial. Unlike the three previous revolutions, it involved a modest change in the energy budget of a numerically, relatively inconsequential eukaryotic species. But by opening up a new uniquely powerful capacity, it led, rapidly on a geological timescale, to a massive transformation in the Earth's biological and physical balance.

Primates first appear in the fossil record some 55 mys to 60 mys ago. Although they have somewhat larger brains and greater mental capacity than many animals, they are only modestly more capable than a number of other creatures. But their presence foreshadowed a series of great revolutions.

The distinctiveness of modern humans, *Homo sapiens*, the descendants of these primates, lies not in their strength or aggressiveness or any metabolic singularity, but in their greater mental capabilities. It lies in their, perhaps I should say 'our', affinity for abstract, creative and symbolic thought, our capacity for reasoning and focused technological development, our ability to make moral judgements and our self-awareness and cognition. However, explaining the acquisition of these capabilities has been a major scientific challenge. Put simply, the pertinent questions in the context of this volume are: what was the evolutionary driver leading to these mental capabilities and how could ancient hominids have afforded to develop these mental capacities when, as we will elaborate, it required the investment of very significant proportion of the energy they acquired from food in their growing brains – food that would, on many occasions, have been in short supply?

Before seeking to answer these questions, it is pertinent to summarise our current understanding of the evolution of *Homo sapiens* (see Figure 4). Let me emphasise, while the broad trend is well understood from research in archaeology, comparative anatomy and molecular biology, new discoveries are being announced regularly and hypotheses accordingly modified. Fortunately, to date, they do not detract from the story I am presenting.

Until about 2 mys to 3 mys ago a number of Australopithecus species (southern apes) were to be been found in Africa [56]. They were largely bipedal but may have also climbed trees. They had small brains of about 380 to 430 cm^3 in volume. Evidence suggests that around this time a smaller species emerged, usually known as *Homo habilis* – the handyman, although there are arguments whether 'habilis' should be regarded as an advanced Australopithecine or the first member of the genus *Homo*. '*Habilis*' was clearly bipedal and had a larger brain than the southern apes at about 600 cm^3. A little under 2 mys ago there is good evidence from the skeletons discovered of the appearance of a clearly '*Homo*' species, namely *Homo erectus*. This species is found from Africa to China and the consensus view suggests that 'erectus' evolved in Africa but then spread quite rapidly to Eurasia. '*Erectus*' had a larger brain of up to ~1000 cm^3 and other anatomical features not that dissimilar to

Figure 4: The human lineage

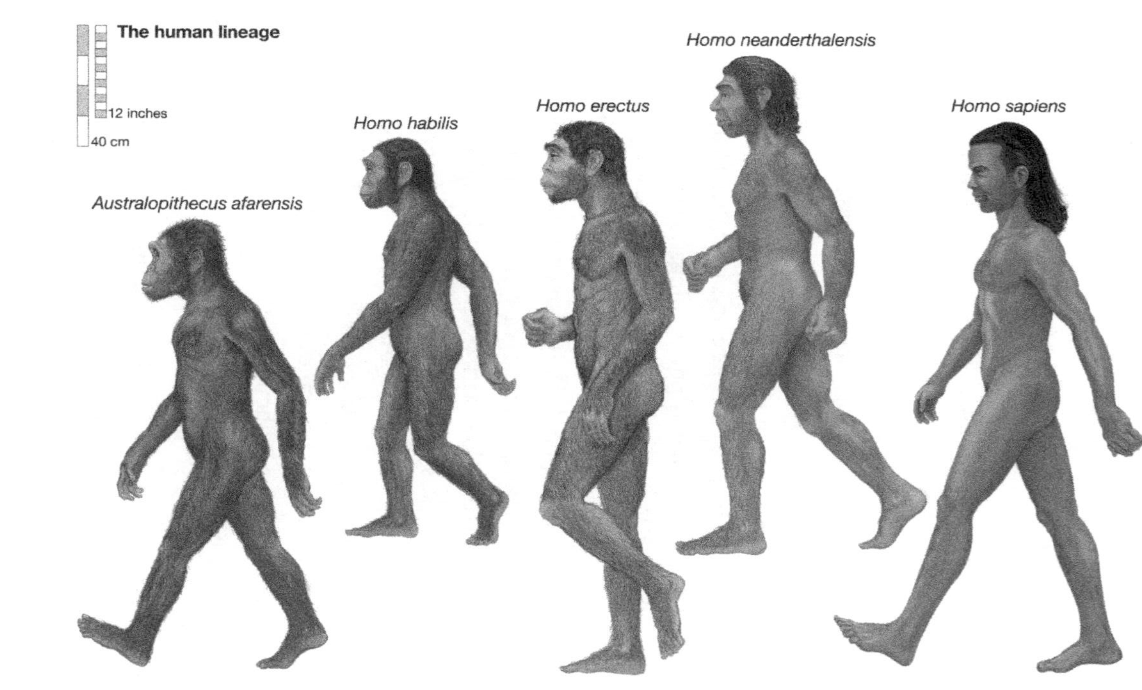

Human lineage (reproduced with permission from *Encyclopædia Britannica*) [55].

Homo sapiens (compare body structures in Figure 4). There is good evidence that 'erectus' lived in organised communities and enjoyed a degree of cultural sophistication. The species was clearly a biped and consequentially had two hands to undertake dextrous work as well as benefiting from the enhanced energy efficiency of running on two legs. The size difference between male and females decreased in this species compared with their ape ancestors, implying major social changes.

It is generally agreed that a number of *Homo* species evolved from *H. erectus*, including us, *Homo sapiens*, and *H. neanderthalensis*, *H. denisova*, and the controversial *H. floresiensis*. Some point to *Homo heidelbergensis* as an intermediate on the road to 'sapiens'. Our detailed knowledge of these species other than humans and neanderthals is sparse and new findings often challenge existing interpretations. The precise timing of the emergence and demise (except for 'sapiens') of these species is much debated. Nevertheless, some facts are undisputed. Only 'sapiens' has survived and prospered. Outside Africa, 'sapiens' carries a small number of genes acquired by interbreeding with neanderthals and denisovians. Up to about 40,000 years ago several *Homo* species co-existed on Earth, although modern humans had arisen at least by 200,000 years, perhaps as long as 350,000 years, ago [57].

The human brain is significantly larger than that of the ancestral species, occupying about 1,450 cm. As an interesting aside, neanderthal brains were likely a little larger than those of modern humans (see note 13). In terms of the body mass to brain weight ratio, the ~1.4 kg human brain exists in a relatively small 75 kg body. There are heavier brains, for example in elephants (running to 5 kg) and in sperm whales (7 kg), but in much larger bodies. There is a broad relationship between body mass and brain mass in mammals but the scaling differs; being nearly log linear in primates excluding the great apes [58] but shallower in other orders. However, size is only a very rough guide to braininess. The number of neurons, their packing and their interconnectedness are more important than simple weight. *H. sapiens* brains have about 86 billion neurons; 16 billion in the cerebral cortex, 69 billion in the cerebellum and some 1 billion in the rest of the brain [58]. Each neuron has a single long extension, an axon, and many small dendritic branches: together they can connect, via synaptic junctions, to as many as 100,000 other neurons. So, there is the capacity in toto to pass signals via perhaps 100 trillion (million-million) synaptic connections; together creating a formidable capacity for data handling (computing power).

The activities of the neurons and their associated glia cells require formidable amounts of energy. This energy is used both to create a membrane potential in the nerve cells, whose controlled collapse is the primary mechanism for sending an electrical signal or impulse speeding along the axon, and to transfer the signals across the synapses between adjacent nerve cells. Of the average minimum total daily energy requirement of modern humans of around 2,000 food calories (kcals or Cals), we use about 500 food calories per day to energise our brains. That is we, as *Homo sapiens*, devote some 25% of our daily energy requirement to our brains. Significantly, the supply of sugar and energy to our brains is also selectively protected in times of calorific stress. This energy economy is strikingly different to that found in other vertebrates including other primates who invest 10% or less of all energy/calorific intake in their brains [58][59].

It has been estimated that a 30 billion neuron brain is, energetically, the maximum affordable by an average 75 kg 'person' living within the limits imposed by the early pre-*Homo*, hunter-gatherer lifestyle. It is postulated that there is a brain-limiting trade-off between the time invested and the calories captured by a largely vegetarian hunter-gatherer living in the style of a southern ape. It appears that day length, resource distribution and digestive inefficiency do not permit this apish lifestyle to support a large sophisticated brain, given the fundamental requirements of survival and reproduction. The proto-human, *H. habilis* (~2.2 to ~1.5 mys) in all likelihood had a somewhat improved diet and might have sustained an estimated 40 billion to 50 billion-neuron brain. But, as noted, *H. habilis* was replaced by the very modern looking *H. erectus*, beginning some 1.8 and 1.7 million years ago.

The question I posed earlier revolved around the relationship between energy capture and energy investment in an expanding and more capable brain. This now resolves into asking what changes could have occurred to allow an early hominid to make the additional energy investment and how could this have created a decisive evolutionary advantage.

Richard Wrangham [59] has postulated that the crucial change in the evolutionary trail leading to large-brained hominids is that *H. erectus* embarked on a 'cooking revolution'. From some 1.5 million years ago there is limited evidence of cooking by *H. erectus*. The earliest observations are, not unexpectedly, contested but later evidence is unambiguous. Cooking of course presupposes an ability to control and use fire. In the context of this hypothesis – the energy sequence is the conversion of chemical energy in wood and twigs (derived from recent photosynthesis) into heat energy. This is then used, by cooking, to make the energy embedded in the food, collected laboriously in many cases, more digestible and palatable and of greater energetic value to that species. Hominid evolution shows these additional calories were invested in brain tissue and brainpower and in launching 'erectus' on a virtuous, self-reinforcing road to greater ability; an ability that resulted in an improved diet, better life and greater reproductive success. This was the evolutionary route from brawn and belly to brain and braininess.

The critical nutritional issue was not simply the calorific value of the raw food (as might be measured chemically, be it vegetables, grains and fruits or meat). Critically important is the digestibility, palatability, safety (the consumption of a number of uncooked 'foods' that otherwise can be toxic), storage, ease of eating and chewing, and possibly even flavour of the food. Cooking can increase the nutritional value of food from about 50% (when raw) to about 95% of the crude 'laboratory' calorific value. The laboratory technique, by fully oxidising the putative foodstuff, assumes that all the energy can be retrieved by the eater. This may be misleading, especially so in the absence of cooking. Therefore, in areas and times when vegetables, fruits, nuts and game were not abundant (which must have been a common experience of early hominids), cooking initiated a huge energetic and nutritional change.

In skeletal shape, *H. erectus* was much like us in having a smaller mouth, weaker jaws and teeth, and a smaller stomach and intestines than those of ancestral species. These features, while compatible with chewing and digesting cooked food, are quite unlike those of the Australopithecines and modern apes, such as chimpanzees and gorillas, who rely on uncooked, largely vegetarian foods. Such animals are obliged to allocate a quite remarkable proportion of their energy to their digestive tracts in order to achieve a modest net gain in calories and other nutrients. So, more belly and less brain! The culinary triumph for one *Homo* band was ultimately a disaster for others, leading to the extinction of all the other Australopithecine and other hominid competitors. Even the surviving great apes are now classed as 'endangered' species by the 'winning' species.

Wrangham [59] acknowledges that, although the major change in brain size is coincident with the appearance of *H. erectus* and cooking, there may have been some earlier influence of meat eating. He accepts that the earlier increase in brain size from Australopithecus species to *habilines* may have been related to a greater access to uncooked meat (despite its low digestibility). Wrangham prefers the name '*habilines*' to *H. habilis*, as, in his judgement, this species doesn't fit neatly into either the genus Australopithecus or *Homo*. His hypothesis has been strengthened by the work of Suzana Herculano-Houzel, as summarised in her book *The Human Advantage* [58]. She has shown that the human brain in size and structure is effectively a 'scaled-up' version of smaller primate brains. The 'neural outliers' are in fact the great apes whose brains are relatively small for their body mass having followed the 'brawn-belly' evolutionary pathway. However she has also recognised what she refers to as the 'primate advantage'. Within the primates the scaling up of neural numbers is nearly linear in relation to brain size. In contrast, in most other animals, larger brains mean larger neurons and an inflationary increase in brain size with a concomitant loss of efficiency. In effect the 'human advantage' has been superimposed on a pre-existing, much older 'primate advantage'.

While I am emphasising the energy chain of 'fire – cooking – calories – brain power', these emerging changes must have stimulated and been reinforced by major social changes. A need and responsibility must have arisen not only for co-ordinated hunting and gathering but also for group cooking, food storage and protection. (A more detailed consideration of the issues from rather different perspectives can be found in Scheidel [60] and Diamond [61].) The ability of a group to organise, co-exist and communicate would have conferred an evolutionary advantage and surely would have become critical. Wrangham postulates that cooking was in all likelihood a largely female responsibility, reinforcing the gender relationships based on the biology of female childbearing and breastfeeding and was an important contributant to the evolution of traditional gender relations and interdependence.

Higher brain power supported the acquisition of more and possibly better quality food and an emerging social order, which in turn precipitated a greater investment in neurons and an enriched social life. Over many thousands of years, this latter became a critical factor leading to the modern 'wired-up', dominant *H. sapiens*. While the cooking revolution took thousands of years to impact on the body structure, brain and lifestyle of *H. erectus*, its full socio-cultural implications only materialised over the next 2 million years.

As I mentioned earlier, with the evolution of biological communities and biomes supporting a variety of animals and plants, new regulatory homeostatic challenges would have emerged. I have argued that, much as homeostatic mechanisms are essential to ensuring the stability, reproductive potential and well being of a single cell, the same principle holds true at the level of multicellular, whole organisms. Indeed, as I will recount later, it was at the whole-animal level that the concept was first proposed. However, this analysis can be extended to the next level of complexity, namely organisms living in communities and in biomes. Homeostatic mechanisms must underpin the stability, wellbeing and success of the genotypes and the communities and groups within which they exist if the individuals are to enjoy reproductive success. The individual must interact with others of the same species as well as with other species. Each must compete to secure resources and mates and respond to external threats and opportunities but also live together. This may appear a far cry from the ability of a prokaryotic cell to avoid heat stress, or to move towards a food store or light, but the principle is similar, as will be elaborated. When social and sentient organisms are considered, the requirement for homeostasis to stabilise complexity assumes yet another form. Social and behavioural responses and patterns, especially but not exclusively in hominids, become vital to the success of the species.

Recently Daniel Everett [62] in his book *How Language Began: The Story of Humanity's Greatest Invention* has proposed another exciting dimension to the *Homo erectus* revolution and to the growth in social interaction. He suggests, in marked contrast to the hypothesis espoused by Noam Chomsky and his colleagues [63], that the emergence of *H. erectus* coincided with and catalysed the emergence of speech. The additional brain power generated by the investment of food energy in neuronal computing must have primed better communications within family groups and tribes. According to Everett, this opened the way to the creation of early, probably relatively primitive, language. He produces various strands of evidence to support his contention, including the emergence of a sophisticated stone technology and a range of serviceable tools (the Acheulean toolkit). Amazingly, *H. erectus* spread throughout much of Africa and Asia, most of the eastern hemisphere, crossing rivers and seas in the process. Everett regards language as a gradually emerging and evolving, but essential, social skill. He concludes that the achievements of *H. erectus* would not have been possible without some language to facilitate and lubricate co-operation and co-ordination. It is clearly tempting to favour Everett's hypothesis and see the fourth revolution as encompassing both the capacity for early speech and other intellectual achievements. Given that we are, and in all likelihood our earlier ancestors were, social creatures and storytellers, it is easy to image the selective pressure around a hunter-gatherer hearth to exchange tales and experiences and that such skills would have conferred physical, social, sexual and evolutionary advantages.

H. erectus, and later *H. sapiens,* are not the only species that cooked their food. Evidence suggests that our close cousin *Homo neanderthalensis* also did so, although this skill did not ensure their survival. I am aware of no evidence about nutritional and dietary regimes of other *Homo* species such *floresiensis* or *denisova* or the recently described *nafeli*. But given the historic relationship between body shape, brain power and cooking, i.e. investing in brain not belly and brawn, it is a reasonable supposition that, given their anatomical features, they also had similar diets and appetites.

The evolutionary pathway followed by *H. erectus* led to *H. sapiens* and, over about 1.5 to 2 million years, to global domination by the latter. However, this evolutionary pathway could well have been quite tortuous. Darwinian competition and natural selection would have continued unabated between the bands of *Homo erectus*, and between 'erectus' and other related hominid species. In this long period, social factors such as the ability to control fire for cooking, social co-operation and co-ordination to acquire food and water and repel predators and enemies, and an ability to transmit information and experience within the group would surely have crucially influenced biological success. From this, *H. sapiens* emerged in Africa possibly as early as 350,000 years ago [57].

Until recently there was a broad consensus, derived from molecular genetic time clocks and the physical anthropology, that *Homo sapiens* first emerged in East Africa about 200,000 years ago. Studies on the accumulation of genetic changes in mitochondrial DNA, which is inherited exclusive down the female line, have pointed to this timeline. This evidence was the source of the 'Eve hypothesis' [64], suggesting that all humankind are the descendants of a very few females in East Africa at about that date. However, as I have mentioned, very recent discoveries in Morocco [57] appear to indicate a significantly earlier date for the emergence of *Homo sapiens* and that to focus exclusively on an East African origin may be mistaken. No doubt more data and additional hypotheses and speculations will emerge in time.

H. sapiens, characteristically inquisitive and competitive as well as co-operative and social, sometimes violent and deceitful as well as considerate and compassionate, set out on a path where more energy was transformed into material complexity and social structures. While the development of secondary material structures such as dwellings is not unique to humans, the scale of this work is: over time, the primate and human advantages of Herculano-Houzel were revolutionary.

There is some dispute as to when *H. sapiens* spread out of Africa, but by about 60,000 to 70,000 years ago modern humans can be found in much of Eurasia. The earliest African human period covered two global glaciations as well as a short warmer interglacial. Early humans had to survive the significant challenges of climate change. During this period the better cold-adapted neanderthals had become established in much of Eurasia. As I have mentioned there is also evidence of a third *Homo* species in Asia, the little known denisovians. Traces of them have been found in Siberia, identified by the unique DNA signature from a few samples. However, genetic mapping shows that both the denisovians and neanderthals interbred with humans and with each other and have contributed their genes to modern humans outside Africa. Neanderthals appear to have died out some 40,000 years ago as 'sapiens' expanded but some of their genes live on in us. The mystery of *Homo floresiensis*, often termed 'hobbits' because of their small stature, remains. Their skeletons have been found on the island of Flores in modern Indonesia. Some suggest this species may have survived until as recently as 12,000 years ago, but the interpretation is hotly disputed.

There must be every expectation that as more data appear, further twists and turns in this complex and tortuous evolutionary tale will emerge. But they are unlikely to undermine the basic story.

Adopting a cautious approach, it is likely that by 30,000 years ago *H. sapiens* reigned supreme as the only advanced hominid on Earth.

These events outlined in this chapter cover several hundreds of thousands of years – a long time in human perception but still a mere blink in geological time (see timeline in Figure 1). During much of this period, it appears only minor changes occurred in the human energy budget and lifestyle. The energy required to undertake all tasks came from food, with some supplementary help from the use of fire and tools. All depended on current, or at least, very recent photosynthesis and on solar irradiance. In all likelihood life in such societies was precarious, competitive but relatively egalitarian. The slow changes in culture and in population imply that human life was harsh and often under threat. High birth rates were likely balanced by high death rates. Some bands of pre-humans and humans would have died out in periods of hardship brought by droughts, floods or cold or warfare. In his book *The World Until Yesterday* [61], Jared Diamond, while discussing modern hunter-gatherer-cookers and subsistence agriculture-dependent bands and tribes, emphasises the critical importance of food supply and the prevalence of hunger in such communities. One inevitable consequence was a premium on defending any additional resources available to any one group. Such concerns may well have dominated the lives of our distant ancestors.

The arrival of *H. sapiens* in many parts of the globe coincided with the extinction of many of the larger animal species, perhaps ill adapted to cope with a clever, hungry predator, but perhaps also burdened by climatic changes. Nevertheless, a quasi-equilibrium in human history appears to have been established for several tens of thousands of years, albeit punctuated by episodic glaciation and other geological and climatic threats. By modern standards, population growth and cultural and societal changes were exceedingly slow but possibly masking a vital evolutionary change. Wilkins, Wrangham and Fitch [65] have suggested that, just as animal domestication was critical to the agricultural revolution (see chapter VI), physiological changes in *Homo* species, summarised as a 'domestication syndrome', seems also to have occurred in humans. These changes, resulting in more cohesive, pliable people, were vital in allowing co-existence and a greater degree of co-operation.

In the previous chapters I have described homeostasis as a mechanism essential to maintaining the integrity and wellbeing of cells and organisms. However, in the new proto human and human worlds of cognition, language, social complexity and material possessions, clearly new regulatory challenges emerged. The need arose for mechanisms to ensure the survival and wellbeing of the individuals and their communities, as well as the cells and organs from which they are composed; that is a step to a higher level on the homeostatic ladder. I will return to these issues in chapter VIII.

In terms of the underlying hypothesis, the fourth revolution can be summarised as follows: by the exploitation and control of one energy supply, namely the burning of wood for fire and heat, early pre-humans found a more efficient way to energise their own work, which in turn expanded their food-energy supplies, possibly also accessing more digestible nutrients. This allowed, cumulatively, a significantly greater investment of energy and matter in neurons and brain power and permitted early humans to work more effectively, partly by increasingly sophisticated tool making. Hunter-gathering became more efficient. Possibly competition amongst hominids was very intense, but survival also depended on cooperation. Gradually, over hundreds of thousands of years the material and social prospects of humans changed. Among the victims of human expansion and predation appear to be some of the mega-fauna that mankind encountered as 'we' expanded over the six inhabitable continents.

Competition and bloodshed, already features of eukaryotic life, as Lane highlights, could well have become more intense as the human population as well as brain power increased in favoured areas.

Comparatively speaking, these changes occurred remarkably quickly, in a little over a million, not a billion or hundreds of millions, years. However it took two further revolutions for these processes to reach their apogee.

CHAPTER VI

'Food Glorious Food'?

Around 12,000 years ago, with the last Ice Age in retreat and a global population of *Homo sapiens* estimated to be around 1 million (dominantly in west Asia, India and China), a new and decisive energy revolution was set in motion. It was the development of settled agriculture and the domestication of specific plant and animal species [5][66][67]. Perhaps it is better characterised as a series of revolutions as similar events occurred independently in several locations around the world within a coeval time frame. The development of crop and animal husbandry allowed a more efficient and expanded conversion of photosynthetic energy into human food and other valued products. By creating larger and more reliable energy sources, human populations of a greater size and density could be sustained. Further, the practice of agriculture required settled communities. Thus villages, towns and later cities emerged, a change that gradually transformed human prospects and social life. As was noted in relation to the ecological food chain, agriculture promoted the acquisition of chemical energy embedded in food materials. This energy was converted into an enhanced ability of more and more humans, and their domesticated animals, to do more and more work that in turn created more complex social and material structures.

Harari [67] calls the agricultural revolution 'history's biggest fraud', painting an unremittingly negative picture of the enslavement of humanity by agriculture. He portrays humankind as being manipulated by plants and animals for the latter's own ends. If this were an adequate insight, then one must ask why human societies have independently repeated the error some 10 times from Mesopotamia to China and from the Andes to the highlands of New Guinea. If true, it implies a poor return on a million-year investment in intelligence.

A more nuanced view is presented by James Scott [66] in analysing the relationship between the beginnings of settled human food production and the origins of states: the latter emerging as hierarchical, often oppressive human constructs. He emphasises the advantages afforded to the small bands of hunter-gatherers in fertile areas, for example in the foothills of Mesopotamia, by the benign conditions that prevailed as the last Ice Age receded. This led to a gradual evolution in human behaviour. It changed from merely harvesting the natural wild grains, recalling that unleavened bread from durum wheat has been dated to some 13,000 years ago (before the agricultural revolution), to seeking to control and improve that harvest. However, Scott suggests that they did not turn their backs completely on the old life. Jared Diamond [68] emphasises the natural ecological advantages offered in some regions, especially west Asia and the eastern Mediterranean. Of the fifty-six wild grass species with large nutritious seeds, some thirty-two occur in the latter region and would have provided hunter-gatherers with desirable food sources and an incentive to experiment with domestication. No doubt also the desire for hides, wool and milk and meat was an incentive to tame animals. But few animals have the characteristics required to permit such domestication, so regions such as west Asia with native sheep, goats, pigs and cattle had a head start.

We in western society tend to be most conversant with domestication of the wheats, barley and various legumes around the Fertile Crescent of west Asia and the Levant. Einkorn and emmer wheat were being cultivated as early as 10,000 to 11,000 years ago. I note 'wheats' as one of the first human ventures into biotechnology was the hybridisation of naturally occurring tetraploid 'durum-type' emmer wheat with the wild grass, *Aegilops tauschii*, to create the hexaploid bread wheat some 4,000 years ago [69]. Remarkably, in view of the current concerns about genetic engineering and Frankenstein foods, this early venture into the field of chromosome engineering was critical to the growth of 'western civilisation'.

In the post-Ice Age era, similar events took place in China with the domestication of rice, millet and beans and pigs starting perhaps 8,000 years ago. A little later maize, potatoes, beans and llamas and guinea pigs were the raw material for the new agriculture in the Nanchoc valley in the Andes and the Oaxaca region in modern Mexico. Sorghum was domesticated in the Sahel. Wild cattle were possibly tamed independently in India and the middle east. In all likelihood, dogs had become human companions and workmates much earlier, perhaps as long as 33,000 years ago while *H. sapiens* was still a nomadic hunter-gatherer-cooker. But the taming of horses and camels came significantly later, ~3,500 BCE.

Some 4,000 years elapsed between the earliest domestication of plants and animals and the rise of the states and cities as hierarchical controlling bodies. This latter change was likely associated with a move to the valley floors, and often with the development of irrigation schemes both in the west Asia/middle east and China. Scott postulates that this change allowed the assumption of power by an elite through taxation and the regulation and allocation of water.

Of course, capturing more photosynthetic energy, through irrigated agriculture, so feeding more humans, plus a contribution from animal labour, allowed more work to be accomplished. Critically, perhaps even inevitably, the control of that work and resultant power was harnessed by elites. In all likelihood, the diets of the poor were grain-based, monotonous, high in carbohydrates but low in some essential amino acids and other nutrients. Studies on human skeletons suggest this diet was inferior to that of hunter-gatherers, certainly in times of prosperity. Why then did humans adopt this lifestyle? Perhaps such a concentration of calorific and social energy and human power in the new urban centres was a temptation to the surrounding hunter-gatherer nomadic tribes. Certainly, it both justified and necessitated the protection (and subjugation) offered by a powerful state, by an elite and by a warrior class.

Although the great majority in these urban settlements may not have been much, if any, better off than in many hunter-gatherers, one can speculate that there must have been perceived advantages. Possibly greater year-long food security was a factor, but maybe the prospect of self-improvement and advancement for individuals and their families was seductive. Maybe greater protection and certainly the scope for a much richer social engagement was appealing. Compared to the emerging civil cultures, the life of the hunter-gatherer would not only have been fragile but monotonous and socially restricted. Living in larger social groups must have offered new social and commercial opportunities, despite also allowing diseases to spread much more quickly. It would have encouraged personal ambition as well as imposing deference. Importantly, these changes would have occurred slowly with many intermediary lifestyles. Maybe humans were drawn into settled communities by the inexorable logic of more food energy priming growing numbers so catalysing a demand for and an expansion in cultivation – a self-catalysing demand? Ian Morris [70] refers, rather disparagingly, to us and our ancestors, as 'Lazy, greedy, frightened people, who rarely know what they are doing, looking for easier more profitable ways to do things' (t. 28). The agricultural revolution seems to fit this characterisation. Maybe it and the development of city-states evolved rather haphazardly; each step a response to an immediate need or opportunity with the overall pattern of urbanisation and emerging power relations not becoming apparent until ingrained. Behavioural psychology has characterised humans as addicted to 'fast thinking' [71], to optimistic superficial intuitive responses, strongly influenced by many biases and heuristics, and to be poor assessors of risk (see also chapter XI). Possibly these traits were important determinants of urban development. However, in addition to Morris's strictures, I would add our ingenuity, sociability and curiosity and lust for power (see also chapter XII) as important drivers on the path to urbanisation.

More food calories supported not only a growth in the population and urbanisation but the emergence of specialisation within society. Urbanisation likely reinforced an existing tendency in hunter-gatherers for the greater resources to elicit both more acute territorial demarcation and greater inequality. As I have noted, by making marauding attractive, it created a need to protect precious assets from such marauders and catalysed the emergence of military castes and heroes. The additional resources allowed the emergence of priestly, administrative and ruling classes and, critically, catalysed the need for new methods for the allocation of resources and of mediating human contacts.

The development of urban life and states clearly offered new regulatory and social challenges that will be discussed in chapter 8 in terms of a wider interpretation of the phenomenon of homeostasis. One of the most far-reaching of these innovations was record keeping. Urbanised and complex societies came to depend on written documents. These allow much more information to be transmitted both within and between generations, and offer opportunities for co-ordination and control. Since writing was a skill confined to the elite for many centuries, it also reinforced social hierarchy. It is tempting to see mathematics and writing as the prime social fruits of the agricultural revolution, much as it is postulated that language and speech were necessary innovations arising from the previous 'fire-cooking-brain power' revolution.

Control and regulation were usually exercised by kings (infrequently queens) or emperors, often deriving their authority from the 'gods'. The gods conveniently consolidated the power of the elite. Early agriculture would have been labour-intensive, partly co-operative and partly coercive and requiring a significant degree of co-ordination and organisation. This must have been especially true when the urban settlements expanded and become dependent on irrigation in the great river valleys of Mesopotamia, China and Egypt. These changes paved the way to mega-enterprises such as the building of great temples. In this world, religious beliefs and practices must have provided a vital glue in communities and a hope for individual escape from oppression. It offered renewal and rebirth possibly in another life or another place, even to the lowest strata of society. Such beliefs held out the promise of future justice and reward as well as cementing support for the elite and promoting social coherence. Without agriculture and its attendant innovations, the great Egyptian, Indian and Persian temples would not have been built nor the terracotta warriors entombed nor the first great literary feats accomplished and retained. In unlettered, nomadic bands, the genius of a hunter-gatherer Homer or Confucius would have been lost to future generations.

The power generated by the additional energy was as capable of being used for destruction as for construction. In the pre-urbanised society, Genghis Khan, or other such figures known or lost to history, could well have existed and flourished but within much more limited realms and a more constrained slaughter. Such is the paradox of the agricultural energy revolution, and indeed the paradox of modern development – brutality and beauty are intertwined.

Solar energy was converted not only into food energy for humans and animals but into non-food crops and commodities such as linen, cotton, silk, hides, wool and furs. These, in turn, stimulated commerce and trade and innovation as well as regulation and legal codes. Nevertheless, all this activity can be traced back to, and was dependent on, photosynthesis and agriculture. There was a limited, albeit vital, ability to store the produce (as is seen in the biblical tale of Moses in Egypt) and to transport food and other items, but the seasonal cycles remained dominant.

In the competition between empires, peoples and charismatic leaders, there are clear parallels with the Darwinian competition after the eukaryotic revolution and with the early competitive hominid evolution. Sedentary pastoral and arable agriculture created the new wealth and culture, and also the hardship and oppression in and between the great competitive empires. The logic of energy permitting work and power made conquest and the forceful acquisition of land and of slaves and dependants an attractive option.

While more food energy allowed work and generated power, often concentrated in a few hands, it also catalysed improved technologies and required more complex and sophisticated social systems. It must be emphasised that even in modern human terms, this was a slow evolutionary 'revolution'. The use of fire and cooking must have taken tens of thousands of years to mature and spread from the small innovative *H. erectus* group to *H. sapiens*' global adventure. Although significantly more rapid, some thousands of years separate the wheat and corn fields of Europe and the USA or the rice paddies of Java from their ancestral progenitors in Syria and Iraq or in China.

Societies whose lifestyles resemble early hunter-gatherers have persisted in remote areas to today. These hint at the ways of our ancestor before the fifth revolution. Modern anthropological studies suggest that where resources, especially food, are limited, either seasonally or continuously, and where populations are small and of low density, such bands are highly egalitarian. Usually living in small co-operative and interdependent groups, they share resources and ranges and live in relative peace with neighbouring bands. Survival in the hard times may have depended on co-operation both within and between bands, although life expectancy was likely low. However, as Jared Diamond [61] discusses, when the resource base improves and the population increases, bands are motivated to be more protective of their assets and to use some of their extra manpower to establish clear territorial boundaries. Both now and in the past such groups may well have been engaged in costly, almost continual, warfare.

This interpretation has its implications for all agricultural societies, including my native Wales. The hypothesis implies that photosynthetic energy yield would be expected to be decisive in the work and power that could be achieved in a given location and, consequently, the focus of the ambitions of a controlling hierarchy, subject, of course, to their management skills, their political values and ruthlessness. The yield and power was derived from the Sun, although also dependent on the fertility of the soil and other geographical factors. Unsurprisingly the great estates, kingdoms and empires emerged in the agriculturally favoured areas. In a local context, the great Marcher Lords were endowed with much greater agricultural resources than the princes of hilly, rainy, mainly infertile Wales. Small wonder the former fought to gain control of the better land. I sometimes wonder what would have been the fate of Britain and Wales if 'Cantre'r Gwaelod' had in fact been a fertile plain stretching from Strumble Head to Bardsey/Ynys Enlli, with easy distance of Ireland and on the western seaway. Indeed, one wonders if the rapid Norman conquest of much of fertile Ireland, while Wales survived, after a fashion, for another 200 years, was not an example of a medieval energy-based, cost-benefit analysis? Was *Pura Wallia* in terms of yield potential in the eleventh and twelfth centuries worth a large and expensive military investment?

In summary, the agricultural revolution, while not only recasting human society and catalysing new technologies, allowed humans to spread, dominate and exploit vast tracts of the world for their herds and crops. In doing so they changed the global ecology. Enormous forests, including those of these islands, extensive native grasslands, and many wild animals had their habitats reduced, and some disappeared. Generally, biodiversity and natural habitats were lost or greatly reduced. Together, fire, cooking, agriculture and urbanisation created this new world remarkably quickly, compared to the previous revolutions that we have been discussing (see Figure 1). Pockets of the old survived but *Homo sapiens* spread to all the inhabitable continents and many remote islands. Mankind has left a mark virtually everywhere. Tribes of hunter-gatherers and subsistence farmers living in inaccessible tropical forests and some other marginal areas survived for thousands of years. But the new societies became ever more stratified. Bureaucracy, formal laws and authority reinforced by some legal or quasi-legal infrastructure became necessities; these were, I will argue, emergent 'homeostatic' regulatory systems (see chapter VIII). In many instances competition was heightened and equality declined. Material complexity also led to greater social complexity as well as demand for goods, be it gold or silver, spices, silks, medicine, jewels or pottery. This gave full scope to the human propensity to affirm status by ostentatious display both in this world and in the grave. All the associated activities and work and power were dependent on the annual or, at best, short-term photosynthetic yield (at the most a few tens or hundreds of years, if we allow for animal breeding or for tree growth in the equation). Humans were using, some might say appropriating, an ever-increasing proportion of the regional and global net primary production; an issue to which I will return.

CHAPTER VII

Fossil Fuels – An Energy Bonanza

'what the world desires to have – power'

The agricultural revolution and the subsequent agriculture-dependent economies of the Neolithic set in train the cultural and technical changes that led ultimately to the sixth climactic revolution. This shattering event overcame life's age-old dependence on neo-current photosynthesis. Mankind found ways of accessing and exploiting hundreds of millions of years of geologically stored, modified photosynthate – fossil hydrocarbon fuels – so catalysing the Industrial Revolution.

Over some 10,000 years, remarkably sophisticated but still essentially agrarian societies gradually accreted technologies, material possessions and novel ideas and philosophies. To supplement human and animal work, wind and water energy were harnessed to power processes such as grinding and milling, but on a modest scale. Timber was not only used as a fuel and for simple tools but also in the construction of sophisticated buildings, ships and other modes of transport. Cloth and garments were woven from wool, linen and silk. Ancient stone masonry of wondrous architectural beauty is a lasting tribute to the skill and endeavour of our ancestors.

An important technological and political driver was the need for irrigation to ensure good yields of the staple high calorie, carbohydrate-rich crops: rice, wheat and maize. These were essential to feed the people, so ensuring biological and cultural continuity and a degree of social cohesion, as well as promoting the greater glory and power for the emperors. By the first century CE, various water-wheel and wind technologies had been developed in the Chinese, Roman and Persian empires. A spectacular achievement of the Roman world was the building of a 62 km aqueduct to power the 16 water mills (the mills of Barbegal) that provided the city of Arles (*Gallulla Roma*) and possibly the Roman navy with flour. In Rome herself, animal power was harnessed to drive rotary querns for mass-producing bread.

Despite the hiatus of Rome's collapse, the Domesday Book of 1086 recorded nearly 6,000 mills in England. Hugh Thomas [72] notes there were some 60,000 water mills in France and perhaps half a million in all Europe by the eve of the Industrial Revolution. These mills could generate power in the range of 5 to 10 horsepower (HP), i.e. 3.6 to 7.3 kW of energy. A 10 HP mill operating eight hours a day for an average working year might have provided around 10,000 MWh of power per year.

By the Middle Ages, some impressively sophisticated technologies had been developed. Indeed, in Kaifeng in the China of the Song dynasty, as early as 1000 CE, coal was used extensively in iron smelting, while in late medieval London, coal was being burnt for warmth [72]. By the 1400s, Chinese and European ships under sail were undertaking great oceanic voyages of discovery. Vaclav Smil

[5] offers a fascinating account of the technical improvements in sailing vessels after the Roman period.

The Pacific islanders, in flimsy craft, made astounding voyages using only wind, tide and human effort to explore and settle islands, from Formosa to Easter Island, not reaching New Zealand until the thirteenth century CE. In contrast, the more accessible Australia was likely first colonised by *Homo sapiens* some 65,000 years ago.

Notwithstanding this growing sophistication and daring and the exploitation of what we would now refer to as 'renewable energy' (see note 14), another great energy revolution would soon transform the world. This was the advent of steam engines driven by the burning of coal. This revolution was still ultimately an exploitation of planetary photosynthesis, but humanity learned to exploit the residual fossil hydrocarbons accumulated in geological sediments over hundreds of millions of years. So, rather than relying on, and being constrained by, the products of recent photosynthesis, there was an energy bonanza – at least for a minority. Importantly, this not only allowed a great deal of work to be carried out but placed mobile, flexible and controllable power in human hands. While it is common to anthropomorphise steam engines, especially in children's books, they are certainly more biddable and less likely to revolt or die than slaves or serfs.

While some early engines did run on wood, the greater energy density of coal gave the latter a crucial advantage. The combination of scientific inquisitiveness, technological innovation and dynamic capitalism in seventeenth-century western Europe rewrote not just the structure of energy consumption but our global social, intellectual and economic assumptions. Energy began to be converted into material and social complexity at an unprecedented rate.

There are many accounts of the Industrial Revolution (see, for example [73]), so in this volume I will limit myself to a very few pertinent aspects. Following Newcomen's prototype steam engine in 1712, James Watt produced a much-improved engine in the 1770s. Fittingly, his name is immortalised as the scientific unit of power. Coal-fired steam engines powered a great revolution by increasing the efficiency and reliability of many processes. Cotton looms and gins were now some forty- to fiftyfold more efficient than when dependent on human labour. New revolution drove the smelting, manufacture and use of copper, iron and, later, steel. In 1824 Aspdin patented the making of Portland cement, leading to the wide use of concrete, so revolutionising building but again with a high-energy demand. The new technologies made transport by train, ocean liner and tramp steamer commonplace: all powered initially by cheap coal.

It is perhaps less appreciated that there was a parallel and equally significant growth in food production during the Industrial Revolution. New agronomic and animal husbandry techniques were introduced. Breeding for higher productivity was initiated and the gradual mechanisation of agriculture allowed a growing population to be fed (see Figure 5). The first combine harvesters were developed in the USA in the 1830s. Much depended in the early days on the import of guano and saltpetre from Chile to maintain fertility. Later the Haber-Bosch process was developed by 1910 for fixing atmospheric nitrogen into valuable chemical fertiliser, but using about 2% of all the human energy demand [74]. These events were transformative and changed the energetic balance in agriculture throughout the western world. Amazingly, by the late 1960s it is estimated that US corn production was more dependent on energy inputs from hydrocarbon fossil fuels than on sunshine! Mackay [4] quotes Albert Bartlett: 'Modern agriculture is the use of land to convert petroleum into food' (p. 76).

The changes wrought by the energy-powered industrial and the second food revolutions were quite extraordinary and the rapidity

Figure 5: Growth in global population and energy use since 1800

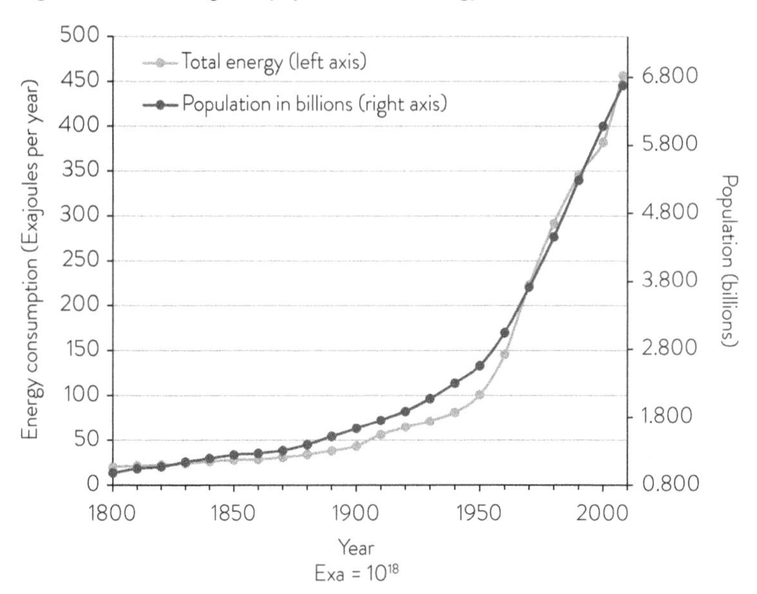

See *https://www.e-education.psu.edu/earth104/node/1347* [75].

of their spread dwarfs all the previous revolutions. The revolution on which humankind embarked is crystallised in a comment in 1776 by the entrepreneur and backer of James Watt, Matthew Boulton. In explaining the new steam engine to James Boswell, he said: 'I sell here, Sir, what all the world desires to have – power' (note 15). Although Watt's original engine only generated 10 HP, it was soon eclipsed by much more powerful versions. A new world was empowered and transformed and the vast reserves of chemical energy embedded in the lithosphere released to do our bidding!

As in other energy revolutions, this empowerment favoured some more than others. The original agricultural revolution offered a substantial political and military gain to an elite. So too industrial power. While globally there was a vast increase in total and per capita energy demand and in economic and population growth [75]

[76][77] (see Figures 5, 6 and 7), these overall trends mask huge inequalities [60]. The trailblazers were the first and greatest beneficiaries. As the revolution spread first within Europe and then in the USA, these countries were given an enormous economic and political boost, enabling them to colonise and exploit others. As before the primary beneficiaries were an elite, quite often the descendants of a landed elite, but even the working class benefited in these countries. However, the peoples of Africa, South America, native North Americans and Australians, and much of Asia, including the once powerful Muslim world of west and central Asia, became very much the poor relations. While the initial colonial expansion

Figure 6: World energy consumption: changing priorities

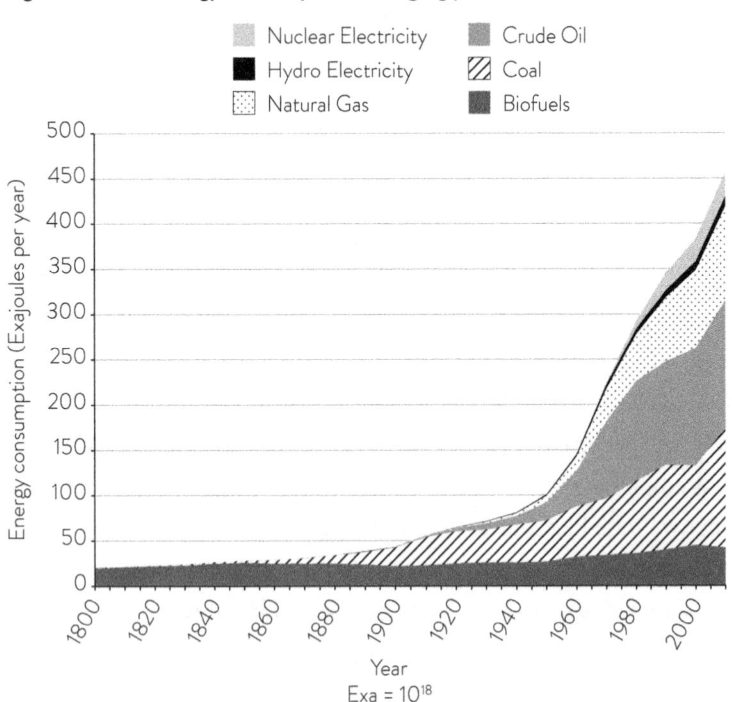

See *https://ourfiniteworld.com/2012/03/12/world-energy-consumption-since-1820-in-charts* [78].

into New England and Quebec and by the Spanish into Central and South America preceded the Industrial Revolution, in most instances its full reach was dependent on the industrial firepower (real and metaphoric) derived from an energy revolution. It is not too far-fetched to view international relations as the product of energy supply and demand and the consequential power relations.

Over time there was a major shift in the fuel preference; an early demand for wood soon gave way to coal, then to oil, and more recently to gas (see Figure 6). In turn, these shifts influenced the political power relations around the globe.

As human energy consumption increased exponentially, so did wealth as well as numbers (see Figures 5, 6 and 7). From a global population of less than 1 billion in the late 1700s (itself nearly a thousandfold increase on an estimated 1 million as the onset of the

Figure 7: Estimation of growth in global wealth

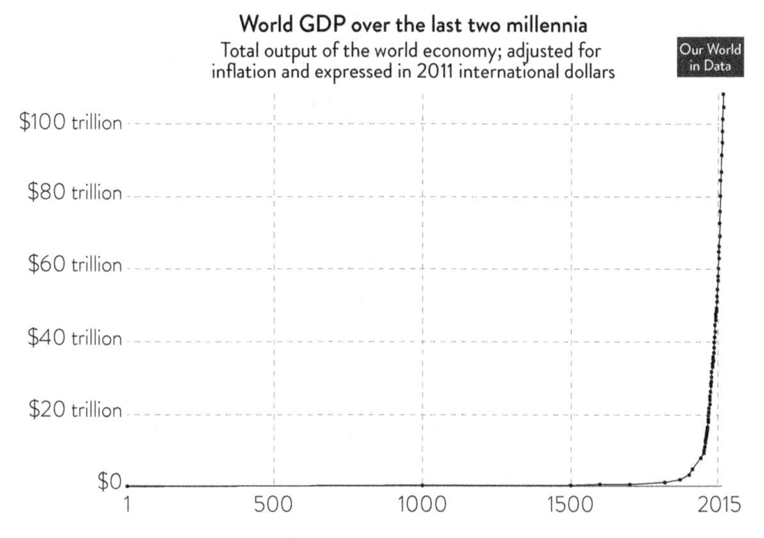

See [76][77] and also *https://www.google.co.uk/search?q=world+gdp+history+chart&tbm=isch&tbo=u&source=univ&sa=X&ved=2ahUKEwjTwfqxkL7eAhUDaVAKHdyiCsgQsAR6BAgFEAE&biw=1920&bih=887#imgrc=nDa6wP58RoifAM:*

agricultural revolution some 12,000 years earlier), human numbers doubled in 120 years, reaching 2 billion in 1923. It has since trebled in less than a century to close to ~7.7 billion today. Estimates of gross domestic product (GDP) show a similar and even more rapid trend as wealth per head as well as total global wealth has increased dramatically [77][78] (see Figure 7) The demand for food and fuel, of course, rose in tandem with the growth in wealth and population. Despite Malthusian warnings of woe, calamity was avoided by huge increases in food production. This occurred partly through improved, fossil fuel-dependent technologies and better husbandry, and partly through the appropriating of vast areas in North America, Australia and Argentina from their native inhabitants by European colonists to grow food for industrialising Europe and the US Atlantic seaboard. Naturally, this rampage was dependent on the imbalance of military might, greatly aided by the explosive power of firearms.

The first signs of a blossoming in a scientific culture of inquiry and experimentation in the seventeenth century preceded the great energetic breakthrough of Watt's steam engine. Indeed, they laid the intellectual and cultural foundation for it and many other practical and intellectual enterprises. The impacts were self-reinforcing as the advent of the Industrial Revolution itself accelerated and gave status to science and technology and to commerce. In energetic terms, few discoveries were of greater significance than that of electricity and later the electron and atomic theory. Although many ancient peoples had some inkling of electrical and magnetic forces, the early experiments of Galvani, Franklin, Volta, Ampère and others led to Michael Faraday's invention of the electric motor by 1821. By the end of the century, J. J. Thompson had identified and characterised the electron, which led within a few decades to atomic theory. Faraday's work presaged our use of electric power that now dominates so many aspects of our lives. Nevertheless, all this activity was dependent on fossil fuels as the primary energy source.

Atomic theory paved the way for atomic power and nuclear fusion and fission. Both enabled catastrophic power to be unleashed in bombs and the latter led to atomic/nuclear power stations. For the first time, the human race accessed energy not derived directly or indirectly from our Sun. In pursuing fusion for electricity production, we are seeking to replicate, in a small way, and control solar processes on Earth. However, the promise of cheap and plentiful nuclear electricity has remained elusive.

The sixth energy revolution empowered the great industrialists of the western world and their countries. The transformation driven by power, technology and capitalism roughly coincided with the publication of the Darwin-Wallace theory of evolution by natural selection and the banner headline of 'nature red in tooth and claw'. The social constructs derived from this theory provided a convenient cover and justification for some of the excesses of capitalism and empire-building colonialism. Inevitably, political power and empire catalysed a growth in bureaucracy and administration and the need for standing armies and navies to enforce the political will and to defend the advantages of elite countries and their favoured sons (occasionally daughters).

Global power political relations remain partly a reflection of the dominant form of hydrocarbon energy. Inevitably, the newly energised states harness both their economic and military technologies and power to influence and exploit, if not necessarily to conquer, and colonise the rest of the world. By the mid-twentieth century the pre-eminence of coal started to wane, with first oil and then gas assuming greater importance (see Figure 6). This power shift benefited counties such as Saudi Arabia, Qatar and their elites. It gave them new status and a capacity to export their fundamentalist, many would claim distorted, version of Islam, much as various brands of Christianity followed conquests of the European powers a century earlier.

The sixth revolution, born of human ingenuity and the exploiting of energy resources laid down over millennia, has been more dramatic, faster and at least as competitive as all its predecessors. In keeping with the patterns of earlier revolutions, while its achievements in material complexity and social interaction and intellectual and cultural achievement has been amazing, so also have been its negative, potentially catastrophic, impacts.

I alluded earlier to the appropriation of the resources of many human communities from the Great Plains and the pampas to Africa to provide raw materials and food for this revolution. Nevertheless, it must be recognised that a significant minority, perhaps as many as 2 billion people, have come by the end of the twentieth century to enjoy unprecedented prosperity. Until recent decades, the main winners were, undoubtedly, the populations, mainly 'white skinned', of Europe and North America. More recently, Japan and latterly China have emerged as industrial powerhouses. Rich elites are now found in most countries. However, a new trend can now be detected, with some of the original winners in the hydrocarbon energy-power bonanza being left behind and becoming embittered and defensive.

The global military threat from the power humans now control is undiminished. The battles of the tribes of New Guinea [61] became world wars and the threat posed by nuclear warhead-carrying intercontinental missiles has replaced that of spears. However, it appears that human nature has changed little and the defence of privilege remains an imperative, but one carrying a global threat.

The fossil fuel-powered Industrial Revolution has created completely new challenges. As was anticipated well over 100 years ago, our technological and economic successes are impacting on the atmospheric, terrestrial and oceanic systems that support planetary life. Atmospheric level of the greenhouse gas CO_2, mainly produced by our burning of fossil fuels, has soared from 260–270 parts per million (ppm) to ~410 ppm in 2018 [79]. This, as shown in Figure 8 [79], is taking our atmosphere to concentration ranges not observed over hundreds of thousands of years and through several Milankovitch cycles. These cycles, arising from subtle changes in the relationship between the Earth and Sun's rotation and progression, are the main drivers of natural global climate change such as the Ice Ages, and are now overwhelmed by human activities. Similar trends are also observed in the atmospheric accumulation for two other major greenhouse gases (GHG) – methane and nitrous oxide – also produced by human activity. The evidence is mounting that

Figure 8: Historic trends in atmospheric CO_2 concentrations

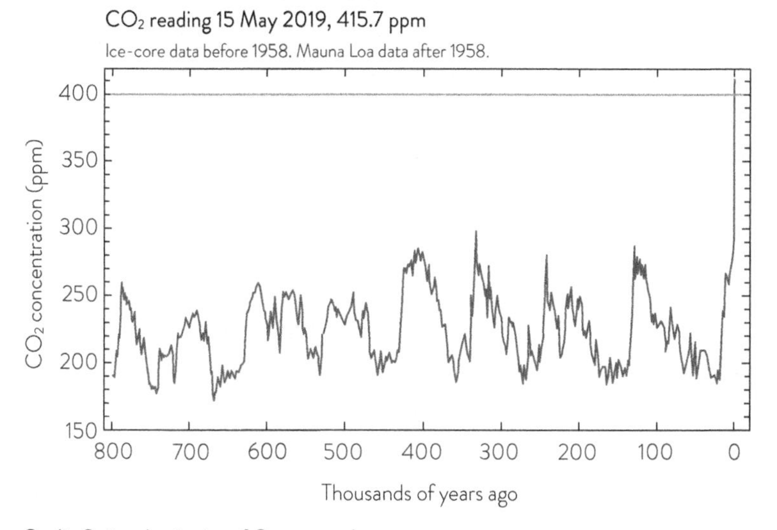

Credit: Scripps Institution of Oceanography.

consequential global warming and climate change is a reality from the Arctic Circle to the Amazonian rainforest. The projections of its unchecked impacts on planet Earth and its human and non-human inhabitants are nothing short of catastrophic.

Although the affluence gained by the sixth revolution is very unevenly distributed, its emotional and aspirational impacts are global. The very success of the sixth revolution is forcing mankind to grapple with an unprecedented set of challenges that I will discuss in greater detail in chapter XI. While acknowledging the critical challenge of anthropogenic climate change, it is important to recognise that human affluence, its needs and numbers have other global impacts. These include greatly elevated atmospheric and terrestrial levels of nitrogenous compounds, e.g. ammonia (NH_3) and nitrogen oxides (NO_x), widespread soil erosion and degradation, loss of habitats, biodiversity and tropic forests, oceanic plastic contamination and acidification, and coral bleaching. Unfortunately, even this list is not exhaustive, but serves to underline the ambiguity of the success of *Homo sapiens'* fossil fuel-fired Industrial Revolution.

The Industrial Revolution and ensuing dynamic complexities have created profound new homeostatic challenges and equally profound dilemmas. What mechanisms have been inherited from our ancestors to help us stabilise these newly emerging complexities? What new mechanisms have we devised? Are they working well? Do they complement each other? What of the fate of other organisms and the biosphere in general? Such questions are highly pertinent to the modern world and to the seventh energy revolution – the move to new energy sources to power our homes, mobility and industry and to provide our food – which do not exacerbate and perpetuate climate change.

How we deal with the impacts of our energy-generating and energy-utilising technologies and the demands of humans on the

Earth's resources and regulatory systems have become the defining issues of our time. I will turn to them in chapters XI and XII after addressing some of the issues and inferences arising from the previous chapters.

CHAPTER VIII

The Homeostatic Hierarchy

The central theme of this volume is the importance of energy fluxes in potentiating work, so allowing the evolution, initially, of highly dynamic but clearly delineated ordered biological structures, in a highly complex entity – a single viable, self-replicating cell. Over several billion years and through five further major step changes in the energy economy, biological and latterly social and material structures of remarkable complexity and sophistication have emerged. As emphasised, the biological structures – single cells and complex organisms and societies alike – must have evolved capabilities to respond to internal and external stimuli so as to maximise their internal 'metabolic' sustainability and, for want of a better term, their wellbeing. These essential regulatory responses are summarised in the word 'homeostasis' (see note 2) [8]. Homeostatic regulation is therefore the bridge joining energy and work to structural stability; without it, the structures would be unstable and life unsustainable. In this chapter, I wish to expand on this basic concept in the light of the first six energy revolutions.

In the case of single-cell prokaryotes, characteristic of revolutions 1 and 2, the situation is conceptually relatively straightforward, although biophysically and biochemically highly sophisticated. There is a direct physical relationship between the cell and its external aqueous environment, so the organism must be able to respond both to external physical stimuli and to internal signals indicating its metabolic wellbeing or otherwise. As new levels of cellular complexity evolved following the third eukaryotic revolution, so new issues came to the fore: initially the relationships between the internal subdivisions within an eukaryotic cell – the organelles – and the whole cell. In the simplest case, such as in single-cell eukaryotic – a protist – the whole cell will still be bathed in its external medium, possibly either fresh or brackish/salty water. Nevertheless, under these circumstances not only must the homeostatic balance between the cell and its environment be maintained but the relationships between the activities within the internal compartments must be co-ordinated and regulated. As previously mentioned, the endosymbiotic assimilation of a bacterium is the direct source of two organelles, mitochondria and chloroplasts, but, as illustrated in Figure 3, the eukaryotic cytoplasm contains both the cytosolic fluid and an array of other membrane-bounded organelles. A further indication of the challenge of co-ordinating the activities of these organelles within a single cell is that mitochondria and chloroplasts have retained some of their own genes independent of those of cell's own nucleus. Obviously, therefore, 'homeostatic' mechanisms must ensure both the internal 'wellbeing' of each of these organelles and the co-ordination of their activities to meet the needs of the whole cell within which they reside.

The organelles are bathed in the cytosolic fluid of parent eukaryotic cell, which is itself remarkably constant in its chemical composition and ionic strength; an example of tight homeostatic regulation around a set point. Comparative biochemistry has shown that the major metabolic pathways such as glycolysis, tri-carboxylic acid cycle and many of the main biosynthetic and degradative pathways are near identical in a wide variety of organisms. As I noted earlier, the conservation of the mechanisms for gene reading and protein synthesis lies at the heart of biology and of genetic engineering. Of course, the conservativeness of cell biology is reflected in the similarity of the DNA in disparate organisms and the old joke that we humans are 50% banana. Overall, there emerges a picture of the conservation of much of the cytoplasmic metabolism, but, in evolutionary terms, whole organisms are nevertheless able to adapt flexibly to external challenges and to internal demands. This balance of conservation and flexibility/diversity has allowed cells to evolve into many types of organisms occupying many ecological niches, although throughout homeostatic mechanisms are geared to maintaining central metabolism within relatively narrow boundaries.

The health of the cell depends on maintaining, with fairly strict limits, sometimes referred to as 'set points', its internal pH, its ionic strength and ion selectivity and the concentrations and fluxes of crucial metabolites. As will be discussed, this principle is applicable at higher organisational levels.

The challenge of homeostatic co-ordination can only increase if we move to consider multi-cellular organisms. At their simplest, such organisms may only exhibit limited cell differentiation and specialisation, but even then another level of complexity arises. Not only must the activities of each organism be regulated and co-ordinated, but this must be superimposed on and integrated into events at the cellular and subcellular levels. As organisms have grown more complex, more highly specialised cells and organs have evolved – to mention only muscle and nerve cells in animals and phloem and stomatal guard cells in plants. In the case of animals, they have also evolved body fluids that act as a bathing fluid for their cells and organs and undertake refined functions, including the transport of messages, nutrients and oxygen and waste products around the organism. A striking innovation was the evolution of nerve cells and nerve junctions to carry and process messages rapidly within an organism; this was of course a crucial precursor to the fourth revolution. The phloem and xylem transport systems have partly analogous roles in plants, while relatively slow electrical signals can also be detected in some instances [47].

It is at the whole animal level that we are most familiar with the phenomenon of homeostasis (see [8] and note 2 for a brief history of the concept). We are accustomed to checking our own blood pressure, heart rates and temperatures and to sending our blood or urine for testing for glucose or electrolyte levels (and a vast array of other tests). Often we are seeking reassurance that we are in good working order; that is, that our homeostatic and metabolic mechanisms are in good shape and maintaining our physical wellbeing. We know also that there is some latitude in the acceptable values and that these may alter with age and other factors, i.e. the set points can vary. Even in the nineteenth century the famous French physiologist Claude Bernard realised the profound significance of these phenomena. He wrote 'La fixité du milieu intérieur est la condition d'une vie libre et indépendante'. In other words, no animal can exist without a healthy, homeostatically regulated functioning physiology.

This is not the place to elaborate on all the intricate physiological and biochemical mechanisms involved. Suffice to re-emphasise that, as structural and material complexity has increased, the homeostatic mechanisms required to sustain each component as well as the whole integrated organism must have increased in parallel. I interpret this as climbing the ladder of a homeostatic hierarchy in conjunction with that of energy and work and material complexity (see Figure 9). Homeostasis has itself an unavoidable energy and material cost. Energy, as ATP and electrical gradients,

Figure 9: The homeostatic hierarchy

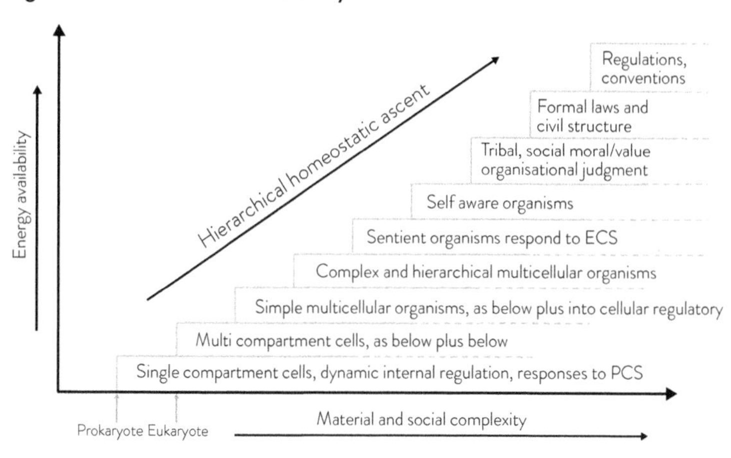

PCS – Physically competent stimuli
ECS – Emotionally competent stimuli

and biochemical mechanisms/structures, must be devoted to the maintenance of the integrity of any cell or multi-cellular organism. In conditions of extreme resource deprivation this may approximate to the 'basal power requirement' of bacteria [80]. But it is not easy to quantify this requirement as it may vary with the metabolic state of the organism. At first sight, it might be anticipated that the resource demands of homeostasis might grow as structural complexity increases. However, since homeostasis is such an integral facet of both cell and organismal biology and there is such tight integration between regulatory processes and metabolism, growth and replication, hard data to confirm this supposition are difficult to find. I will return to this and related dilemmas later.

A revealing illustration of the balance of flexibility and conservation inherent within these systems and its homeostatic implications can be found in plants and algae. Unlike animals and fungi, plants, as I have mentioned, must generate large surface areas. Large leaves allow as much sunlight as possible to be intercepted and consequently to maximise the capture of photosynthetic energy. As we have seen, this energy is used to fix atmospheric CO_2 into sugars and other forms of chemical energy. However, plants, consequently, have to cope with a negative trade-off, as a large surface area can lead to increased water loss by a process called transpiration; this is controlled water evaporation through pores (stomata) on the leaf's surface. This process is essential to drawing nutrients up from the roots to the shoots, but excessive water loss threatens wilting and desiccation. On the other hand, a large root surface area can promote nutrient capture and water uptake by allowing the plant to explore a greater soil volume. In the case of the essential nutrient, phosphorus, many plants enjoy a symbiotic relationship with specific root fungi (termed 'vesicular-arbuscular mycorrhiza' (VAM)) to further increase the effective soil volume that can be exploited. De facto, they trade plant sugars from photosynthesis for fungal phosphate. However, investing large quantities of carbon and nitrogen et cetera in creating large volumes of cytoplasm is inordinately energetically and materially expensive in immobile, sessile organisms. Such demands could themselves limit growth. Plant cells have therefore evolved to have large vacuoles that maximise surface area while limiting the demand of expensive metabolites (see Figure 3(b)).

However, the water in these compartments must equilibrate, i.e. both must exert the same osmotic pressure, otherwise water would rapidly flow from one compartment to the other, resulting in desiccation or bursting of either the cytoplasm or the vacuole. This trade-off between volume and the investment of material and energy cost is accomplished by the vacuoles accumulating a wide range of energetically cheap, easily available inorganic and organic solutes. The precise nature of these solutes depends on the specific ecological niche. My colleague Roger Leigh and I coined the phrase 'the fastidious cytoplasm and promiscuous vacuole' to highlight this dichotomy [3]. That is, the primacy of conservation of the biophysical and biochemical characteristics of the cytoplasm (with its attendant homeostatic mechanisms) is complemented by a flexible use of the resources to assist a species to adapt their volumetric growth in specific locations. It follows that the homeostatic mechanisms for regulating the concentration of contents of the vacuole must differ significantly from those of the cytosol and the other cytoplasmic organelles, but nevertheless must be kept in step. At a higher organisational level, similar principles apply at the whole-plant level and determine the distribution of solutes to flowers, seeds, growing apices and other tissues. Cells within tissues with high rates of metabolic activity, e.g. seed embryos or pollen, are dominantly 'cytoplasmic' in their cytology and chemistry.

Further homeostatic regulatory mechanisms are required when the social as well as the physical complexity of the life cycles of some multi-cellular organisms are considered. In this case, the mechanisms have evolved to be much more than the simple, albeit already sophisticated, responses of single and multi-cellular creatures to 'threats', e.g. the need to escape predators, or to exploit 'opportunities', by swimming towards new food supplies. Nevertheless, they are all recognisably related to each other.

Figure 10 is reproduced from Antonio Damasio's inspirational book *Looking for Spinoza* [20] (p. 45). In this volume, he explores, in the main, homeostatic regulation at levels above those that I have been outlining. He does so with reference to animals, especially humans. As a plant scientist I believe it is important to recognise that comparable mechanisms occur in higher plants and that these basic principles arise directly from complexity, or more precisely the stabilisation of complexity, and are not uniquely 'animal'.

Damasio makes a compelling case that, in order to exist and hopefully to flourish, any animal, including *Homo sapiens*, must have a range of internal regulatory mechanisms to cope with outside stimuli and to integrate the responses at all levels. So far, I have been discussing the reactions of organisms to 'physically competent stimuli' (PCS). However, more advanced animals also respond to 'emotionally competent stimuli' (ECS). These require the presence of a nervous system and a 'brain'. They elicit initially automatic physical metabolic responses, e.g. pupil dilation and/or raised pulse rates, and these reactions then extend to other reflex and immune responses. These are the components of an increasing elaborate homeostatic regulatory system. They have evolved to maintain the integrity and the 'wellbeing' of the organism, not only in a physical context but increasingly in its social and behavioural milieu. Anyone owning a dog or who has been in close proximity to other animals, including elephants and dolphins, realises that responses to ECSs are not confined to hominids. A recent article in the *New Scientist* [81] suggests that even bees, with very small brains, can

Figure 10: Damasio's homeostatic tree

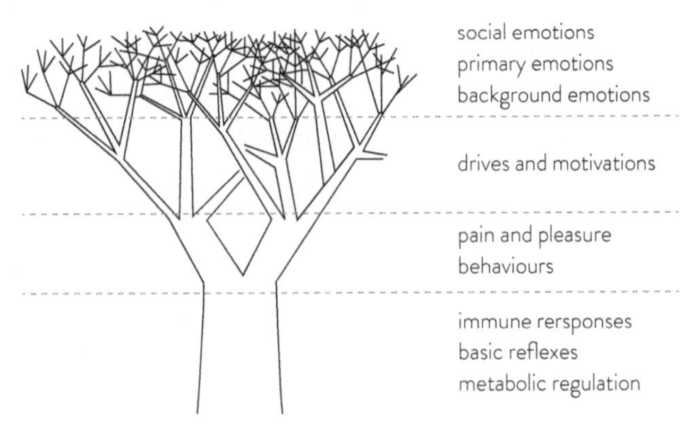

Reproduced with the kind permission of Professor Antonio Damasio [20].

make choices and show reactions that appear to have an 'emotional' content.

Damasio's analysis moves beyond the basic pain/pleasure responses, through drives and motivations, such as hunger and sex, to more sophisticated responses that he divides into background, primary and social emotions. And at the summit of his regulatory tree (see Figure 10) are our feelings. He writes of a 'nested sequence of responses', with the latter being the outermost branches.

In self-aware humans, our total responses bring together not only our initial, largely automatic, physiological/emotional responses to a current stimulus (be it a PCS or ECS) but also recollections of previous events and an expectation of their consequences. These neural responses are superimposed on a picture of our total body state carried in our brains, together with this anticipation of future possible reactions, based partly on our experiences and preconceptions. Together they generate and define our feelings and drive our decisions and actions. This integration lies at the very heart of our humanity and our culture.

I cannot do justice here to the richness of Damasio's analysis and his interpretations leading from neurology through to ethics and culture. His recent volume *The Strange Order of Things: Life, Feeling and the Making of Cultures* [82] explores these latter aspects in some detail. He offers a definition of homeostasis as regulating life 'within a range that is not just compatible with survival but also is conducive to flourishing, to the projection of life into the future of an organism or a species' (p. 25). I would again emphasise that homeostasis should be seen as the essential bridge binding the structural complexities derived from energy, work and power to their stability and longevity. I contend that a homeostatic ladder (see Figures 9 and 10), climbing from the responses of unicellular prokaryotes, eukaryotes and multi-cellular organisms right through to those of social animals including humans, can be identified. Of course, this concept carries us a long way from the simple homeostatic responses of a bacterium. However, it provides an evolutionary interpretation of how we interact and achieve some contentment/ wellbeing within our own tribe and how we might cohabit with the rest of the animate and inanimate world.

As Damasio has pointed out [20], our social life as adults requires much more than automated 'homeostatic' emotional responses. It requires a capacity to control and modify these underlying 'automated' solutions and an ability to modulate our basic emotions and feelings. I will return to the issues implicit in these non- or partly automated rapid responses in the context of the seventh energy revolution and climate change in a later chapter.

In Table 3 below I reproduce Damasio's summary of the social emotions and their consequences (p. 156 in case).

These emotions contribute to the raw materials for our feelings and our drives and our interactions at a personal, community and wider national and international level.

If we place emotions and feelings within the context of the energy revolutions, it is clear that the acquisition of brain power and neurological complexity has not only promoted self-awareness and our cognitive capacities but also created a requirement for new homeostatic mechanisms. The whole hierarchy becomes important to the wellbeing and survival of individuals and their genes and also to that of the whole group. It is within such groups that we as individuals experience shame, contempt, compassion and awe, and, indeed, may live or die as a result.

The fourth revolution set the stage for much more than greater acquisition of material complexity and social differentiation. By about 5,000 years ago humans began to live in large villages and cities. These acquired, indeed required, complex human hierarchies

Table 3: Our social emotions, their sources and consequences

EMBARRASSMENT; SHAME; GUILT	**ECS**: weakness/failure/violation in the individual's own person or behaviour. **Consequences**: prevent punishment by others; restore balance in self, other or group; enforcing of social conventions and rules. **Basis**: fear; sadness; submissive tendencies.
CONTEMPT; INDIGNATION	**ECS**: another individual's violation of norms (purity; co-operation). **Consequences**: punishment of violation; enforcing of social conventions and rules. **Basis**: disgust; anger.
SYMPATHY; COMPASSION	**ECS**: another individual in suffering or need. **Consequences**: comfort, restoration of balance in other or group. **Basis**: attachment; sadness.
AWE/WONDER; ELEVATION; GRATITUDE; PRIDE	**ECS**: recognition in other or self of contribution to co-operation. **Consequences**: reward for co-operation; reinforcing the tendency toward co-operation. **Basis**: happiness.

and social and power structures to sustain their stability. The analysis therefore suggests that the homeostatic tree extends to human constructs such as justice, morality and religion, to our socio-political organisations, our economic theories as well as our sense of personal identity. Such non-automated human devices and reactions should, I suggest, be seen as an elaboration and extension of social 'homeostatic regulatory systems'. They are critical to our capacity to co-exist in relative peace and harmony, in much the same way as the activity of cells must be integrated into a whole organism.

However, any consideration of these human regulatory devices is compromised when accommodated 'within any human construct whose wellbeing is being prioritised'. Even within small groups or tribes, some individuals were in all probability more influential, esteemed and powerful than others. Were they, consequentially, favoured by the custom and 'regulations' of that group? How then did, and does, individual wellbeing relate to the collective wellbeing? Differentiated cells co-work for the wellbeing of the whole organism, but it is scarcely valid to extrapolate this concept uncritically to humanity. Individual or even group wellbeing cannot equate to the collective wellbeing of a society, a nation or humanity in general, or vice versa. I will return to this when discussing the seventh revolution and our ability to respond to the challenge of the anthropogenic climate change.

A complementary analysis has been presented by the Polish psychiatrist, Antoni Kępiński (see [83]). He used the term 'information metabolism' to describe the processes by which the healthy inner order of an organism is maintained by a continuous exchange of information between the organism and its environment. He identified three hierarchical levels for these exchanges, one being biological, the second emotional and the third socio-cultural. This approach clearly resonates with that of Damasio and the one taken in this monograph. The term 'information metabolism' has its attractions for describing the interactions and interdependence of the organised, conserved, low-entropy state of a living being or organism, including humans, and its more randomised higher entropy environment.

Other aspects of the homeostatic ladder, especially after the agricultural and the industrial revolutions, not addressed by Damasio nor, as far I am aware, Kępiński, are the regulatory demands arising directly from the energy-dependent, work- and power-driven technological changes brought about by the sixth revolution. The steam and internal combustion engines did much more than energise trains, cars and trucks. The latter created a demand for roads, traffic regulations, police and courts to apply these regulations, for driving tests and licences, bureaucracy and taxes, in part to pay for the roads and road safety and for hospital beds, nurses and doctors to treat the maimed.

Even this list is incomplete. The pollution from both the engines and tyres created a need to monitor these pollutants, to legislate and to enforce regulations and, in extreme cases, for more prison accommodation. Vast networks of suppliers and manufactures were created. The complexity of the technology stimulated clever tricks to mask the true volume of the toxic emissions and other deceits. Similarly, steam trains, ocean liners and aircraft, all energy-driven innovations, have transformed society and required immensely complex, national and international regulatory and legal frameworks, not just physical infrastructure. While trains brought about integrated timekeeping in Britain, aviation has rewritten international time. It is also the case that, although not referred to as homeostasis or homeodynamics or information metabolism, control systems to ensure the stability of complex systems in response to external turbulence lie at the heart of aeronautics and engineering in general.

These technological innovations also stimulated profound social and cultural change. New personal and collective possibilities and cultural norms emerged. The growth of great cites, industrial towns and suburbia was accelerated. New elites emerged and efforts were made to counter the power of these elites by non-governmental social structures from unions to mutual societies.

Mass cheap transport led to mass sport and tourism. Can one imagine Arsenal playing Liverpool, still less 'Barça' or Celtic, if the teams and their travelling fans had to travel on horseback or a galleon? Can one contemplate weekend breaks in Dubai without cheap air travel? Can one imagine a modern energised consumerist world without mass education, medical care, a raft of environmental and planning regulation and 'health and safety'? All have been driven by the hydrocarbon-powered Industrial Revolution. In many ways, airport terminals are the ultimate temples to modern consumerism as well as the flagships of our technological mastery and energy profligacy. The crucial homeostatic role of air traffic control is only obvious when the comptrollers strike!

As emphasised, new homeostatic regulatory mechanisms must accompany the exploitation of novel energy sources. Thus an ability to couple any such new sources to additional work and power rests on both factors in coordination. It appears that such a capability can not be taken for granted. Even in the case of photosynthesis, only about 1% to 2% of the total solar radiation reaching planet Earth is actually captured, although, of course, solar energy also powers other vital planetary activities (such as the hydrological cycles, and the natural greenhouse phenomenon which brings the mean planetary temperature to a mild, life-compatible ~14°C). Each of the energetic step changes represents a new form of coupling. While it is possible to describe each as a specific event, a deeper understanding of their feasibility or inevitability is as elusive as was the specific phenomenon discussed in relation to the evolution of life itself (pages 15–16). It appears that great energy fluxes, such as those arising from nuclear fusion or fission or major volcanic activity, can not be harnessed directly in this way. Nevertheless, it appears that throughout planetary history, by promoting more work and power, energy fluxes have enabled material and latterly socially complex structures and cultures to emerge. I am arguing that there is a golden thread, itself made possible by work, running from the earliest cell (LUCA; note 4) to modern society. The stability, sustainability and well-being of humans, other life forms and our planet depends on the effectiveness of these mechanisms and the continuity of that thread.

CHAPTER IX

Emergent Patterns

Before considering the challenges posed by our fossil-fuel energised economy, I wish to pause and consider whether any other patterns can be discerned and tentative conclusions drawn from the six revolutions previously discussed (that is, from the emergence of the first viable cell, the capture of the energy from the sun through photosynthesis, the evolution of complex eukaryotic cells, the hominid investment of added energy in brain power and the agricultural and industrial revolutions).

Acceleration

The planetary rate of change has accelerated dramatically, especially after the fourth revolution. The gaps between revolutions have declined from over a billion years to a mere 200 years, with a concomitant increase in the rate at which the changes have radiated out globally. It is worth recalling that, in physics, power is defined as energy used per unit time. So, an enhanced energy supply (or great change in its effective availability or the efficiency of its use) will allow more work to be accomplished per unit time. From this perspective it is unsurprising that a huge increase in available energy and power is associated with a sharp acceleration in the rate of change. After the human energy revolutions, it seems likely the rate of change was itself subject to a feed-forward reaction stimulated by population growth. Provided the resources allowed such growth (but see 'Ceilings' below), more individuals could do more work (usually hard physical labour; see Figure 2) and provide more power (usually to an elite) to create further social and material complexity. Since population growth can be quite rapid under certain conditions, this might in and of itself have gradually accelerated the evolution of complexity between the major step changes.

Ceilings

In *The Vital Question* [35] Nick Lane notes that prokaryotes use some 80% of their energy in the protein synthesis required for their own replication. He implies that they have little or no 'free energy' to invest in additional material complexity. Certainly, the geological record suggests that over nearly 2 billion years there was little change in material complexity in the prokaryotic world despite its biochemical prowess. Perhaps under these conditions there was little selective advantage in multi-cellularity and too high an energetic cost. This constraint, Lane suggests, was lifted by the endosymbiotic absorption of mitochondria/bacteria *and*, crucially, the shedding of all but ~13 for the proto-mitochondrial genes. According to Lane, this process dramatically increased the amount of energy available per gene, catalysing the biological transformations previously discussed.

It is illuminating to compare Lane's putative release of eukaryotes from a biological constraint with the hypothesis advanced by Ian Morris in his exciting and revealing volume, *Why The West Rules For Now* [70]. His work examines the pathways by which societies have evolved by comparing historic developments at two major poles. On one hand, he considers east Asia, effectively China, and on the other, initially west Asia and latterly Europe. In the latter, the main focus of social development migrated slowly from Mesopotamia to the eastern and later central Mediterranean in Roman times, then to north-western Europe in the eighteenth and nineteenth centuries, and most recently to the USA.

Why, Morris asks, was the UK able to impose its will on China during the Opium Wars, as opposed to China, with its rich history, large population and technical prowess, making European nations its playthings? In order to answer this question and to make semi-quantitative comparisons between development at the two poles over the past 16,000 years (from 14,000 BCE to the present), Morris has proposes a social development index. This is based on four criteria: (a) energy capture; (b) organisational capacity – for which he uses urbanisation as a proxy; (c) information processing; and (d) the capacity to make war. He recognises a 'great chain of energy' running through human society over this period. Before the Industrial Revolution, some 75% to 90 % of the development score in his index at both poles could be attributed to energy capture. Even as recently as 2000, energy accounted for 20% to 28% of the index.

Although Morris's index extends over the whole period, but I have reproduced in Figure 11 only the data for the period between 1600 BCE and 1900 CE. Unsurprisingly, these overall indices dating back to the agricultural revolution describe curves very similar to those shown in Figures 5 and 7 (energy consumption, population growth and GDP per capita). After an extended period of very gradual change, they explode in a phase of extraordinary exponential growth from the start of the Industrial Revolution. During this latter period the 'West' takes a significant lead. However, the graph in Figure 11 is of particular interest as it illustrates periods of both 'western' and 'eastern' dominance. During these specific periods, both cultures appear to have attained similar developmental plateaux, the former in the Roman period and the latter in the Song dynasty. Both societies then suffered marked retreats from these peaks. Morris goes into some detail about the forces of disruption that afflicted societies after their periods of glory and dominance – often in the form of the four 'Horsemen of Apocalypse' (pestilence, war, famine and death), but, undoubtedly, failure in governance may also have contributed to their demise.

Morris postulates that in an agrarian society there is an upper limit to the socio-economic development that can be achieved, a limit

Figure 11: Social development index to 1900 CE

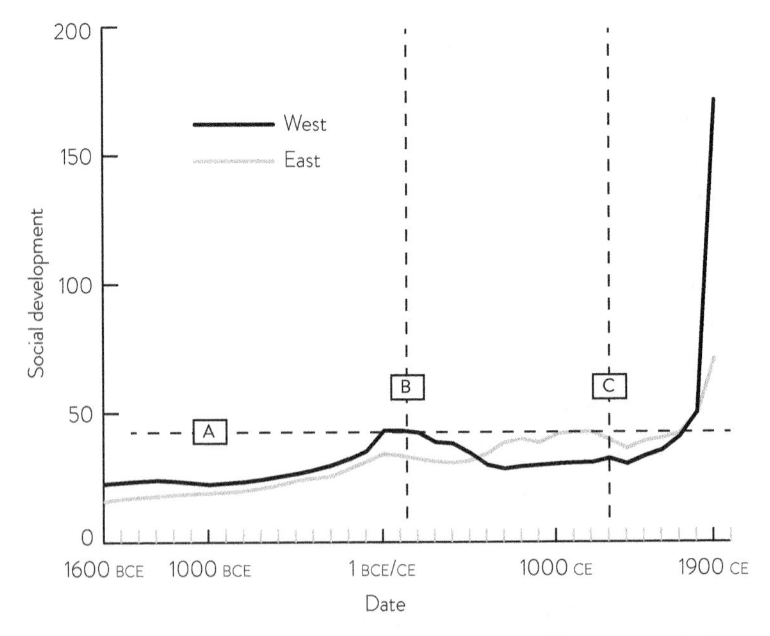

Line A – possible 'agrarian' developmental threshold; Lines B and C show respectively the Roman (western) and Song dynasty (eastern) maxima followed by declines and then the recent impact of the Industrial Revolution.

Reproduced with the kind permission of Professor Ian Morris [70].

imposed by the energetics of the food supply and the absence of an alternative controllable source of power, i.e. a very limited supply of a non-current photosynthetic energy. The Industrial Revolution and the power supplied by Boulton and Watt swept aside this limitation.

The much-quoted work of Earl Cook in the early 1970s [84] expressed related concepts in terms of per capita energy use. He starts his analysis with proto-humans living on some 2,000 food calories a day (kcal per capita per day) at the birth of the fourth revolution. His estimate of 2,000 food calories per day *without cooking* over a million years ago is higher than the more recent postulates of Herculano-Houzel [58] and Wrangham [59]. He then takes a fine-grained approach to later changes, seeking to separate early from later more advanced medieval agriculture, and then distinguishing 'industrial' society (Britain: *c.*1875) from 'technological' society (USA: *c.*1970). He suggests, as seems very reasonable, that, during the period from *H. erectus* (~1.7 million years ago) to some 15,000 years ago, there was a very gradual and modest increasing trend in energy capture to about 2,000 to about 3,000 kcal per day when all aspects of culture, e.g. clothing, shelter, decoration, ceremonials, et cetera, are taken into consideration. These must be conceived as very generalised average numbers as it is virtually certain that, in individual locations and discrete societies, growth, albeit slow, was neither smooth nor progressive. Rather, humanity's timeline was punctuated by disasters, collapses and periods of recovery. But crucially, change before the agricultural revolution was limited and human and hominid societies plateaued for well over a million years.

These patterns imply that, certainly in the case of the eukaryotic, hominid (*H. erectus*), agricultural and industrial revolutions, the emergence of new energy regimes allowed the 'complexity stasis', exhibited for a significant period prior to the step change, to be swept aside and new complexities to emerge. Let me add that I am of course not minimising the importance of many cultural innovations and social changes during these periods, but their material impacts were modest compared with that engendered by the energy revolutions. The concept I am advocating has many similarities to the 'punctuated' biological evolution, as advocated by Stephen Jay Gould [51].

If one seeks to integrate the evidence from molecular and evolutionary biology through anthropology and neurology to the recent agricultural and industrial revolutions, I suggest a remarkable pattern emerges. Not only did each energy revolution catalyse material and social change, but it removed an energetic constraint that had imposed an upper limit on the developmental and evolutionary potential of the previous phase. This can be envisaged as the release of an emergent potential by the acquisition of more energy and new power. Each revolution allowed a new flourishing, although, in many cases, at the expense of creating new losers.

Co-operation and Competition

Obviously, each revolution has created new opportunities both in relation to life forms – simple cells, eukaryotic cells, multicellularity, and hominids – and later human societies, but each has encouraged new types of, possibly, more intense competition. The neo-Darwinian selection and evolutionary pressures found in pro- and eu-karyotes seems to re-occur, with modification, in hominid competition, in competing city-states and empires, and in current capitalism. These competitions have inevitably created winners and losers, although this was possibly less apparent in the earliest prokaryotic phases of lateral gene transfer.

In the case of the second revolution, namely the harvesting of light energy through photosynthesis, I am not aware of any great change in the type of competition. Other than the adaptation of the

H^+ gradient/H^+-ATP synthase from the LUCA to be part of the new 'photosynthetic energy capture apparatus' and the likely acquisition of photosystems I and II from different bacteria, there is little obvious change in material complexity or competition. Nevertheless, by releasing prokaryotes from any energetic dependency on abiotically produced, reduced compounds, this change was indeed revolutionary. It allowed not only increased prokaryotic biodiversity, but, in time, the colonisation of land. Lenton and Watson [10] emphasise the long-term influence of the oxygenation of our planet on both its biology and geology.

The balance and relationship between competition and co-operation is itself a matter of fierce debate. While the common image is of nature 'red in tooth and claw' and is exemplified in the motif of the 'selfish gene', I have referred to several examples of co-operation and symbiosis, including the critical and singular step of the emergence of eukaryotes. In the case of mitochondrial endosymbiosis, it appears that the bacterial donor paid a price in the loss of most of its genes, while a small subsample of the donor genes prospered and multiplied in the female line of countless species.

It is worth emphasising that in biology there are many examples of co-existence and co-dependence, a coming together to create a greater whole. I have alluded to lichen as symbionts of fungi and micro-eukaryotic algae and cyanobacteria. Most corals are also derived from a symbiotic growth of invertebrates – cnidarian – and microalgae. Nitrogen fixation in higher plants arises from symbiosis between bacteria in the root nodules of legumes and the plants themselves. Indeed, the plants pay a price in yield potential for investing their photosynthate in the activities of the azotobacter, but from which they in turn gain fixed nitrogen.

Ruminant animals, whose complex stomachs are fermentation vats of bacteria, archaea and fungi, depend on this co-operative relationship, but of course it arose in the context of competitive selection. From a ruminant's perspective, their microbial fermentation sack allows them to make effective use of grasses and browse and so to exploit the vegetation over vast areas of the Earth's hilly and semi-arid surface. From a microbial perspective, the rumen is a desirable habitat with a reliable supply of nutrients. Added to this, since the agricultural revolution, humans have exploited this relationship to produce a convenient source of nutrients, meat and milk and so on. In turn, humans ensure the reproductive success of their ruminants and provide them with short-term protection but an early death. There are now some 1.5 billion cattle and 1 billion sheep on this planet; the chain of co-operative exploitation has been extended. One unanticipated negative impact of this co-operation, as will be discussed in relation to the seventh revolution, is that these animals are now an important source of GHG emissions, although they are the bedrock of many ancient and some human cultures.

Other non-ruminant grazing animals, as diverse as horses, elephants and kangaroos, although lacking specialised rumens, all depend on microbial fermentation in their intestines to extract energy from their fodder. Microbial fermentation is relatively unimportant to human nutrition, as we depend mainly on cooked and easily digested foods, although fermentation supplies us with our yogurt, beer, cheese and wine. Even so, we carry within us about as many prokaryotic cells as our own genetic inheritance of eukaryotic cells, and our wellbeing appears strongly influenced by this microflora. Evidence is growing that our flourishing, to use Damasio's terminology, requires a well-adapted gut flora whose total genome greatly exceeds our modest ~23,000 genes. Perhaps, when we see a demagogue in full rhetorical flow on our screens, we should be aware that he can be truly described as a receptacle for fermentative micro-organisms.

These are but small examples of a greater truth, exemplified by the great global nutrient cycles of nitrogen, oxygen, carbon and sulphur and the workings of the biosphere. Life on Earth is interdependent, not only in relation to the energy chain and the role of bees and others insects in fertilisation and many other phenomena, but also in relation to all the cycles of all the chemicals from which complexity is built.

As we move to the fourth and fifth revolutions, the roles of co-operation and job allocation and differentiation are increasingly obvious in relation to the exploitation of personal human mental energy and agriculture-dependent communal energy. Survival and success required a mixture of co-operation and competition, an ability to collaborate and to compete, to lead and be led, to fight and to flee. There must have been a premium of working for the common good, even if it required groups to abandon the old or the infirm to their fates in crises. Almost certainly there was an expectation that young men would fight to the death to protect their group from wild animals and at times rival tribes and to die as heroes for doing so. It seems probable that such a society would have favoured fairness, loyalty and bravery, as well as organisational and leadership abilities, in order to bring in a steady food supply and to live in relative safety. Cowardice, cheating and insubordination would have been viewed harshly and there would have been a premium on beliefs and rituals to maximise family and group commonality and cohesion. Studies such as those of Diamond [61] show that such societies continued to exist until very recently, well beyond the sixth revolution. Unsurprisingly, behavioural psychology shows that these traits still exercise a strong influence on our behaviour in modern industrial society.

The great wealth of eukaryotic diversity, including the tortuous paths of hominid evolution, of course bears witness to the importance of competition. A new dynamic was released when sexual reproduction, through meiosis and crossing over, promoted the exchange of genes carried on the male and female chromosomes. This mechanism added a new dimension to the genetic variability on which natural selection could operate. It supplemented and sharpened that which prokaryotes derived from mutations and from inter-species gene transfer and exchange. The fundamental biological model of natural selection rests at the gene level, although the actual selection must be at the level of the adult-reproductive phenotype. There is a continuing controversy about the possibility of selection at the group level [62].

Biological competition must have remained crucial after the fourth revolution. It seems improbable that the small isolated groups of *Homo erectus* and later *Homo sapiens* did not differ in the their ability to find and cook food, and so to invest in better brains and enjoy greater reproductive success. However, at this stage a new factor was emerging – the ability to communicate, to teach and transmit new skills to the next generation and not exclusively to their own offspring – and with it the concept of a collective or tribal good. Even if the group was of limited genetic diversity, this was a fundamental change. Tribal information transfer meant competitive selection became more than an exercise in genetic selection. On the evidence of modern anthropology, it appears that finding a mate from outside the very tight family was part of the lifestyle so reduced the risk of inbreeding while increasing the means of dispersing new knowledge.

Competition remains at the heart of modern human society, although no longer expressed through classical natural selection. With the advent of the small states after the agricultural revolution, competition took on new and sometimes very damaging forms. The city-states, nations and empires, large and small, competed and fought for ascendency, for loot and trade, and for the control of energy – food and labour. This has continued to this day, catalysed by human ambition and greed, and by ethnic, cultural and religious identity. As I mentioned, there are suggestions that these developments had been preceded not only by the domestication of crops and animals but of humans themselves (see [65]). This change, it is claimed, is written in our genes and our physical features. Pinker's analysis [86] also suggests that not only did early humans become more domesticated but that violence has gradually decreased over time. Again, there appears to be a paradox: competition, sometimes violent, has driven societal change, but humans have been domesticated – become more docile – in order to co-exist.

In modern society, competition takes on many forms, some based on ideas, ideologies and dogmas and belief systems as much as on economic and military might. Some of these can be related to fundamental aspects of human behaviour arising from the homeostatic hierarchy and our automated emotions and feelings, others not. As I will discuss, these latter traits are open to manipulation and, at times, being traduced. The popularisation of Darwinian competition, natural selection and the survival of the fittest has had a profound impact on human society and its norms. Social Darwinism [87] has been used to justify eugenics, fascism, colonial domination and exploitation and the status of a capitalist elite. One of Social Darwinism's earliest advocates, Herbert Spencer [87] coined the phrase 'the survival of the fittest'. However the co-operative strand within biology has received less popular and political attention. With the emergence of the market economy and latterly globalisation, as we are reminded daily by politicians, commercial competition has become a global driver. I will consider some of the implication of this emphasis in more detail when discussing the human response to the seventh revolution.

Information

In addition to the energetic aspects already discussed, a parallel series of revolutions in information transfer and processing have occurred. As is described in a number of famous books [12][13][14][16], the DNA triplet code in our genes was, from billions of years, the method by which information was transmitted from one generation to the next. With the arrival of proto-humans and humans, the methods of transfer took a further leap; first, by simple speech and later the ability to incorporate this knowledge in myths and sagas that could be shared within communities and could span generations. Even in the neanderthal world it appears that the visual arts were important in conveying images and inspiration. The formalisation of letters, words and grammar, and mathematical symbols after the agricultural revolution, enormously expanded the human capacity to transmit both specific 'how-to-do' information and our cultural mores and values.

The advent of the printing press a few centuries before the Industrial Revolution greatly increased the speed, spread and impact of the world of books and manuscripts. While the transmission of data and inspiration by writing and calculation remains unchecked, in the last decades it has become digitised. The amount of information at our disposal has soared and the speed of its transmission increased more than a thousandfold. Some would claim we now suffer for 'information overload'.

Over the millennia, the ability to do work and to utilise power, and its transmission of a DNA blueprint to achieve this, has been a major determinant in the evolution of life on this planet. But this may be changing. Information in other forms may be becoming a driver. In chapter I, I alluded to the concept of entropy, suggesting it is loosely associated with a system's molecular disorder and I referred to Schrodinger's now discredited but important idea of life being a system for accumulating negative entropy, i.e. 'order' at the expense of increasing the disorder or entropy of its environment [39]. More recent theoretical work suggests that order is better thought of as a function of entropy divided by the information capacity of the system; information itself being defined as the resolution of uncertainty. The Boltzmann definition of entropy (see note 1) revolves around the number of microstates of the molecules that are consistent with the overall macro-state of the system, e.g. its temperature, pressure, or volume as a gas. Thus, entropy can be interpreted in terms of increasing or decreasing the uncertainty and hence the informational content in the system. These are very challenging and, to an extent, contended concepts – sufficient in this discussion to recognise that there exist close relationships between energy fluxes and their exploitation and changes in the entropy and the informational content of physical and biological systems.

Caveats

It would be misleading to overplay these broad generalisations. There are important global events such as the long lag before the emergence of the first eukaryote and the advent of eukaryotic multicellularity, the division into the plant and animal kingdoms, the appearance of *H. sapiens* (possibly some 300,000 years ago) and the culture changes apparent some 40,000 years ago, and many, many others, which cannot be accounted for by the six energy revolutions outlined. Similarly, since the agricultural revolution, culture, politics, religious, ethical and social values, and economic norms and concepts have had an enormous influence. Early in this volume I noted that the relationships between energy availability, work done or power employed gave no insight into the ethics of that activity; that caveat deserves to be re-emphasised.

It is abundantly clear that important specific changes, some related to the energy economy, have occurred outside the revolutions I have highlighted. These are important but gradual evolutionary changes between the step changes. Examples include the development of hominid bipedism that increased the efficiency of hominid motion, and the colonisation of land by plants, which increased the capacity of global photosynthetic energy capture. Also, as is delightfully illustrated by Smil [5], prior to the Industrial Revolution, the evolutionary technological changes in the exploitation of wind to power shipping and trade were vital. Consequently, it took almost a century for emergent steam power to out-compete sail on the ocean, as exemplified in the 1870s by the great tea cutters, the *Cutty Sark* and the *Thermopylae*. But in the end, it did so with dramatic consequences. It is an interesting exercise to try and reimagine politically, militarily and socially the last 200 years based only on non-fossil fuel energy. Likely the Sun would have set on any emerging British empire much earlier.

In framing this narrative in terms of seven revolutionary or step changes, I am not contending these were single, sudden events. Rather, I see a number of modifications and alterations that have combined over time to bring about a revolutionary change; each of these has reverberated, cumulatively, through time. This approach chimes with that of Lenton and Watson [10] and Maynard Smith and Szathmáry [9], but less so with the views of Smil [5]. He interprets the energy-driven changes over the last 15,000 years as a cumulative evolution. I appreciate his position, but in Figure 12(a) and (b), I present both the original figure from Smil and one redrawn using the same data on an even timescale. I see the data in Figure 12(b), as well as Figures 5 and 6, as fully justifying the

Figure 12: Comparison of energy prime movers on uneven (a) and even (b) timescales (after Smil [5])

(a)

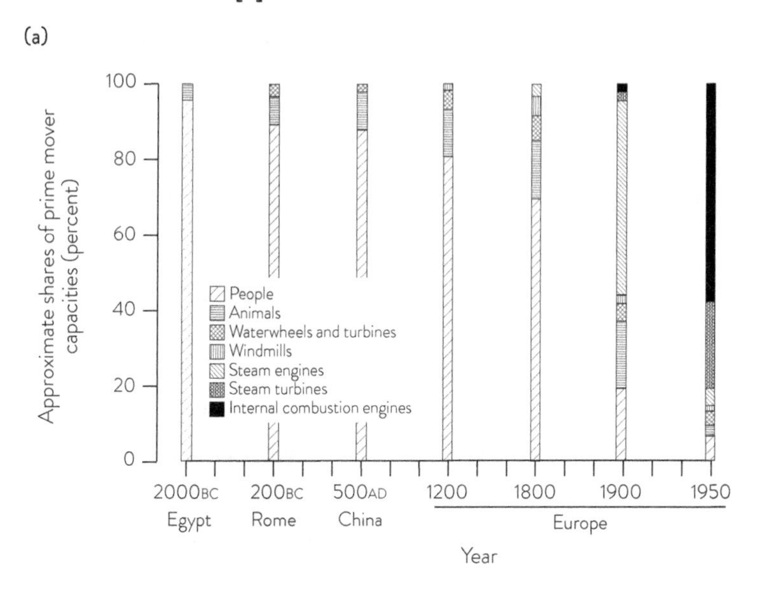

(b)

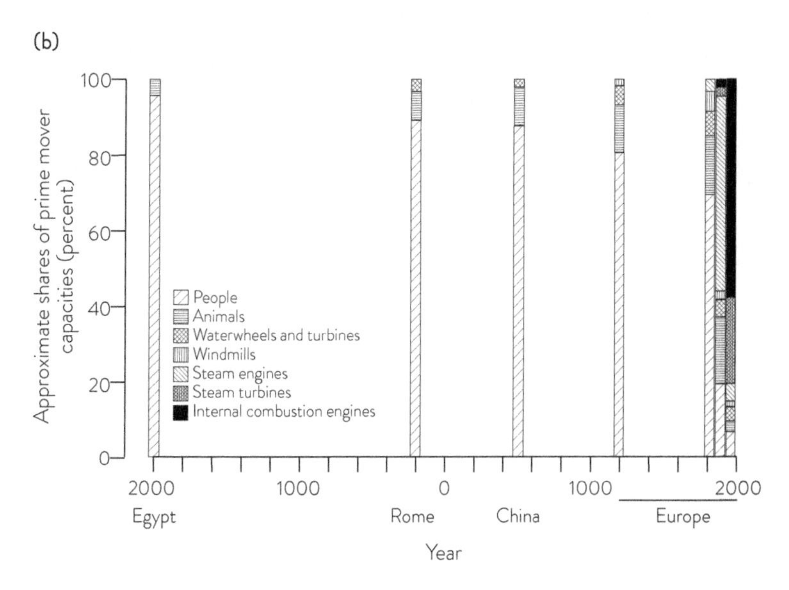

Reproduced with the kind permission of MIT Press [5].

label of the industrial 'revolution'. Who can doubt that the rate of change displayed on an even time frame, with its dependence on fossil fuels, is truly dramatic? Smil's data also underline, as mentioned earlier, the dominance of human and, to a far lesser extent, animal labour to any human endeavour well into the nineteenth century; indeed, even today in some communities.

Human societies were constrained by manual labour, although the implication in the 'man' of manual is misleading. Hard physical work had no gender barrier, albeit often assigning different roles for the males and females. Naturally, such labour has defined the experience of individuals and their societal roles. The rapid demise of this system within a few generations has not only changed gender relations but challenged the basis on which many based their self-worth, as well as our physiology and body weight. Such labour dependency also meant that many great projects, be they the great cathedrals of Europe or the Great Wall of China, have occupied many individuals for very many years. Taken together Morris's development indices [70] (see Figure 11), the exponential growth in global wealth, measured as GDP (see Figure 7), post 1800, and the energy use data clearly demonstrate the revolutionary nature of the sixth step change.

CHAPTER X

The Gathering Storm – Greenhouse Gases: The Effluence of Affluence

Homeo sapiens has emerged globally dominant from these energy revolutions. Paradoxically, this dominance is now both threatening and being threatened. The lifestyles of the affluent and aspirational classes, the main beneficiaries of the sixth revolution, are open to being undermined by the impacts of their own GHG emissions and by climate change and Anthropocene degradation. The lives of the very poor, who have benefited least, are being made even more precarious.

It is important to appreciate the scale of human energy demand in a country such as the United Kingdom. You will recall that in the fourth revolution the investment of some 500 food calories per day (only an additional 0.6 kWh per person per day) in additional neurons and in brain power, beginning only about 2 million years ago was the catalyst. It levered in, after two further energy revolutions, astounding human power, technical prowess and material wealth. The resulting energy demand is nothing short of staggering. In Britain, Mackay [4] has estimated the average individual energy use in 2007 to be some 195 kWh of energy per day, equivalent to

~250,000 kcal/food calories a day. That is a total average individual energy demand a 100 times our basic metabolic requirement and at least 50 times more than that which supported our ancestors. Some over a certain age will remember sitting in front of a single-bar electric fire on winter's night trying to keep warm. For them, this comparison may be useful. A one-bar electric fire uses 24 kWh a day. So, it is as if we have about eight such fires running continuously, day and night, all year long to support our individual lifestyles!

The human metabolic requirement is only about 2.2 kWh per day. Mackay [4], more generously and realistically in the modern world, allowed us 2,600 kcal per person per day – that is 3 kWh each. In the developed world, meeting this basic need requires a substantial investment of energy in farming, fertilisers, agrochemicals and the food supply chain equivalent to about 15 kWh per day per person. This reflects the disparity between our primary food energy requirement and the energy inputs into the agricultural and commercial systems that supply us. However, our primary food-energy budget is dwarfed by the totality of terrestrial and marine transport, space heating and cooling, air travel, industry and just consuming (see also Figures 6 and 7).

Given the huge inequalities in wealth and lifestyle, the energy and consequently CO_2 footprints of the jet-setting elite from any country must be at least double, probably treble, the mean, even in the 'rich' countries. Energy use permeates all aspects of modern life. This, as we have discussed (see also Figures 5 and 12(a) and (b)) [4][5] is supplied largely by burning fossil fuels. (Note that the total GHG emissions from the food chain, which include methane and nitrous oxide, are significantly greater than those due to CO_2 alone. These mainly arise from land use changes, e.g. ploughing land or cutting down forests, as well as transport and agrochemicals.)

It must not be imagined that our dependence on hydrocarbon fossil energy has reduced our dependence on the world's annual photosynthetic resources. In a detailed study, Imhoff and Bounoua [88] concluded that mankind, in the seventeen years from 1981 to 1998, utilised appropriated approximately 20% of the annual global terrestrial net photosynthetic production (NPP) to meet our demands for food (animals and crops), wood products, fibres and other goods. They noted significant regional variation. NPP utilisation varied from about 80% in south-central Asia and a little over 70% in Europe, to only 6% in South America. The global population was under 6 billion when this study was carried out. It is now about 7.7 billion and is not projected to equilibrate until nearing, or maybe exceeding, 10 billion. Consequently, food, fibre and fodder demand can only increase; an increase compounded by the taste of the growing middle class in China and elsewhere for 'western', McDonald's-type, meat-heavy, high-fat, high-sugar diets. Thus, the proportion of the world's NPP claimed by humans seems destined to increase. Although there may be a minor but ecologically distorting offset by anthropogenic atmospheric CO_2 and N (NO_3^- and NH_4^+) fertilisation, it seems certain there will be a catastrophic squeeze on the energy supply chains of many other species.

In a startling recent paper, Bar-On, Phillips and Milo [89] have sought to assess the amount of biomass on earth and its distribution among the various biological kingdoms. They find that terrestrial biomass, dominantly plants, is two orders of magnitude greater than marine biomass and that human biomass is now an order of magnitude greater than that of all wild mammals combined. Humans and their livestock outweigh all vertebrates with the exception of fish. They also estimate that, from the onset of human civilisation, the amount of living matter on Earth has fallen by a half, suggesting that not only are we appropriating more of the Earth's biomass/photosynthate resource but we are actually decreasing the total available to all life.

Officially, the Food and Agriculture Organization of the United Nations (FAO) estimates that some 25% of global GHG emissions arise from food production and related land-use changes, e.g. forest clearance. However, this calculation excludes the embedded carbon in fertilisers and agrochemicals and the energy costs of post farm-gate transport, processing, refrigeration, supermarket sales, et cetera. A figure of more than 30% is a more realistic. Our own simple projections [90], based on an assumed, optimistic cut of 85% in CO_2 emissions from the energy chain by 2040, but a continuing pro rata increase in emissions of N_2O and CH_4 from the food chain and CO_2 from land clearance, suggest that by the mid-century, climate change may well be perpetuated by our food emissions alone.

Humanity's triumph must be tempered by the realisation that, unchecked, the emissions of GHGs from fossil fuel use and food production and our appropriation of other global resources have the capacity to undermine the affluence and wellbeing of rich and poor alike. Our energy use, workload and power are precipitating revolutionary, potentially chaotic, changes in the global energy balance. It is quite possible, therefore, that a fifth horseman will be added to the four traditional Horsemen of the Apocalypse, although probably the traditional agents of disasters will suffice.

To summarise the current situation: the atmospheric concentration of the major anthropogenic greenhouse gas CO_2 has risen nearly 50% from pre-industrial levels of 260 to 270 ppm, to reach well over 410 ppm in 2018 (see Figure 8). Despite the many conferences and governmental promises, global CO_2 levels continue to rise remorselessly (~2 to 3 ppm per year). There are claims from industry and governments that human energy-related CO_2 emissions have plateaued in the last two years. Worryingly, this is not reflected in the atmospheric chemistry. This could be due to the well-attested human capacity for dissembling, but worse it could mean the activation of anticipated feed-forward reactions, e.g. CO_2 release by the oxidation of the tundra/peat or a decline in carbon capture by tropical rainforests. The atmospheric levels of two other GHG, methane (CH_4) and nitrous oxide (N_2O), both partly derived from the food chain as well as the energy economy, also continue to rise. These three gases, together with ozone and chlorofluorocarbons (CFCs), are the main greenhouse gasses causing this planet to retain heat that otherwise would be irradiated back into space. This assertion is supported by the basic laws of physics, by the records of global surface (see Figure 13) and oceanic temperatures (the former extending back over a hundred years), by the continuing melting of glaciers and polar sea ice sheets, by the deep historic

Figure 13: Recent global surface temperature trends

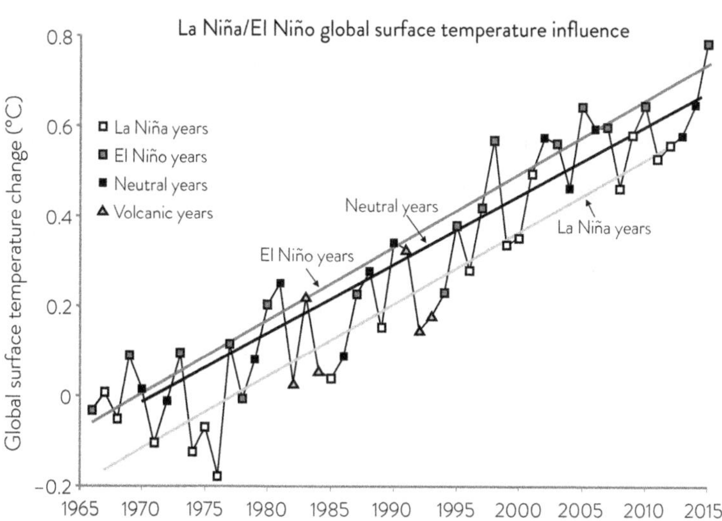

See *https://www.google.co.uk/search?q=el+nino+years+graph&tbm=isch&tbo=u&source=univ&sa=X&ved=2ahUKEwjmj8m41sDeAhWLalAKHcceD1IQsAR6BAgFEAE&biw=1920&bih=887#imgrc=J9H5x7xgwXyiCM* [91].

climate records revealed in ice cores, by dendrochronology and by climate modelling.

The global warming and climate change caused by this extra heat/energy retention is having, and is projected to have, dramatic effects on the biosphere, atmosphere, cryosphere and the oceans [22][23][24]. The gravity of the challenge is such that at the United Nations Framework Convention on Climate Change conference in Paris in December 2015, all nations agreed to cut GHG emissions, with the stated objective of restricting the mean global temperature increase to 2°C, but better to only 1.5°C. Unfortunately, this accord did not include binding commitments to action. In 2016, influenced by a strong El Niño (compare the El Niño line with other years; see Figure 13) the increase reached over 1.2°C, measured as an anomaly from the late nineteenth-century levels, i.e. from the period when the fossil fuel-fired, Industrial Revolution began to take off.

The evidence clearly shows, firstly, that anthropogenic global warming and climate change is already occurring and that further delay in responding to the challenge will be seriously damaging to human society. Earth systems have huge inertia and the embedded changes will unfold inexorably over decades and centuries. Secondly, self-reinforcing, feed-forward reactions pose a major threat as they may accelerate climate change and prime catastrophes such as global sea level rises of many metres threatening major coastal cities. The melting of the glaciers, which provide all-year round water for major conurbations, is a major threat, as are severe fires, droughts and floods. Cuts of more than 67% in global GHG emissions are required in thirty years.

If it is assumed that humans are of equal value, as many religious and ethical leaders have argued over several thousand years, although to little avail, then the most affluent societies and individuals should accept the largest and fastest cuts. It is worth re-emphasising that it is no longer a matter, simply, of country differences but of differences in the energy and carbon footprints of various socio-economic groups; the affluent, be they Chinese or American, Australian or Arab, have the heaviest footprints and largest impact. The poorest billion emit per head perhaps only a twentieth of the GHG emissions of the richest billion and essential food emissions are a greater percentage of their footprint.

There is a tight and damning knot joining global warming, population growth and inequality. The additional CO_2 being added to our atmosphere by human activity equilibrates rapidly but takes hundreds of years to decay naturally as it is only very slowly eliminated from the atmosphere by oceanic and soil processes. This is despite the annual oscillation in the atmosphere of CO_2 of about 8 to 9 ppm due to the annual cycle of growth (photosynthetic carbon gain) and decay (respiration) mainly in the northern hemisphere. Therefore, one can assert with confidence: first, the higher the human population, the lower the annual ration of CO_2 each individual can emit. Secondly, the longer the global community prevaricates and fails to react decisively, the steeper and harsher the cuts required to avoid catastrophic temperature rises (see Figure 15). Without decisive action, parts of the world will become uninhabitable due to a combination of high temperature and humidity, maybe even within fifty years, whilst other areas will be lost to the oceans. Over extended periods, wet bulb temperatures of over 37°C (that is a combination of 100% relative humidity and 37°C) are physiologically damaging to humans; our bodies cannot cope. Such conditions will lead to tracts of the Blue Planet being abandoned and undoubtedly the poor will be the least able to cope or escape.

Poverty and inequality impact on this crisis in several ways. Most obviously, emissions per person are much greater in the richer countries (e.g. USA 20t CO_2; Wales 13t; China 7t; India 4t; Chad 1t;

and a global mean of ~7t per head) [92]. Although in all countries the affluent and well-off are able to command more energy, it is the poor, who generally have higher birth rates and who carry out the more weather-exposed work, that will suffer most directly from the effects of climate change. Nevertheless, many of these pin their hopes for a better life on the current global economic model predicated on continuous economic growth (both total and per capita GDP), and which treats climate change as a tangential externality (see chapter XII).

Historically, there has been a strong link between GDP growth, energy use and increasing GHG emissions. Over the past century this growth has supported a huge increase in population (cf. Figure 5), but, historically, capitalism has also tended to enhance inequity (see, for example, Piketty [93], Atkinson [94], Raworth [95]). Only episodically has this latter trend been reversed, at least temporarily. According to Walter Scheidel [60], significant reversals in inequality are associated with major disasters: wars which mobilise and kill huge populations, or plagues or other catastrophes. He offers copious examples including the Black Death (bubonic plague) in the mid-fourteenth century in Europe and the aftermath of the Great Wars of the twentieth century. Scheidel then sees the embedded tendency to inequality as a force driving investment and economic growth, only countered by massive human disasters such as the Black Death or catastrophic wars. Not a happy scenario!

In the UK our population is ageing. Such an ageing population is inevitable and essential, both locally and globally, if population growth is to be stabilised. But it is seen as a political and economic threat as it places an extra financial burden on the state and on working members of families. In the UK we have resorted to importing cheap labour to provide care for our elderly and other menial work. Such immigration is nevertheless highly unpopular. Increasing tax revenues to compensate is also highly unpopular. Clearly, we are facing, or perhaps more accurately, are refusing to face, a set of unpalatable dilemmas (cf. Figure 14).

In contrast to the annual cycle of the respiratory release (in winter) and photosynthetic capture (in summer) of CO_2, the net removal of anthropogenic CO_2 by geological and oceanic reactions is very slow. As a result, it is our total emissions since the Industrial Revolution really got underway – see Figures 5, 6 and 12 – that will determine the mean global temperature rises that we will subject ourselves to and have to try and cope with.

As mentioned, the countries of the world at the Paris Climate conference aspired to limiting that rise to 1.5°C and committed to one of less than 2°C – but without legal force. We have already used well over 50% of the 'quota' compatible with these objectives, with

Figure 14: Projected carbon emission profiles compatible with 66% chance of mean global surface temperature anomaly of less than 2°C

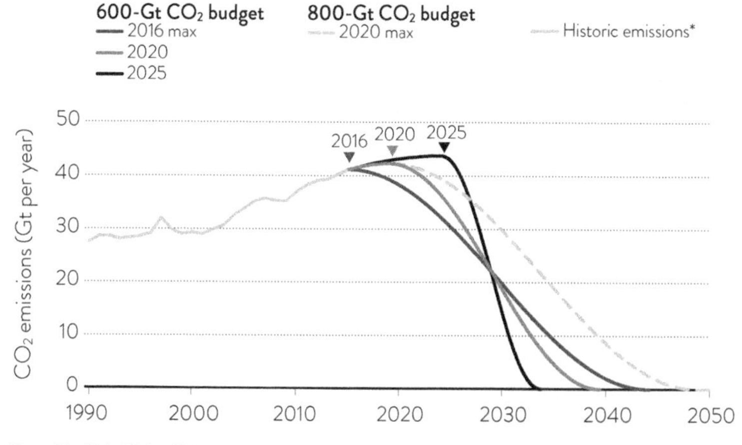

*Data of the Global Carbon Project.

Reproduced with the kind permission of Professor Stefan Rahmstorf [97]. See *https://media.nature.com/original/nature-assets/nclimate/journal/v4/n10/extref/nclimate2384-s1.pdf* the original industrialised counties having taken a disproportional share. The remaining budget depends on the estimations of the global 'equilibrium climate sensitivity', i.e. how much the earth's mean temperature will rise with a doubling of the pre-industrial atmospheric CO_2 level, i.e. to ~550ppm. The consensus value is about 3°C; with some suggesting only 1°C (despite the data in Figure 13), and others favouring higher values, citing loss of the Amazonian rainforest as C sink, oxidation of tundra, changes in the albedo of the Arctic, etc., as additional factors. Figure 14 shows two possible scenarios for the necessary cuts, one more generous than the other, giving a reasonable possibility (66% chance) of avoiding 2°C rise. Critically, both underline the simple but frightening fact that dramatic cuts are required immediately, that any delay makes things worse, and that by mid-century we will have exhausted our carbon budget.

Despite this dire warning, human CO_2 emissions continue to grow, and the rise in the atmospheric CO_2 concentration (which reached 415ppm in early May 2019) is, if anything, accelerating. A number of powerful countries such as the USA and Australia are actively promoting the use of coal, the most polluting of the hydrocarbon fuels. Major countries such as India and China continue to build coal-fired power stations, and companies supported by the banks continue to search for new oil and gas reserves.

Unfortunately, in relation to anthropogenic emission and the seventh revolution, the greater the changes that are required, the greater is both the likely resistance of the better off to any concept of equal shares or to shared responsibility, be that at home or abroad. If we have difficulty finding funds to care for our elderly and infirm, what chance is there that adequate largess can be found for the battalions of poor and climate migrants in or from far-off lands. Paradoxically, the probability is that inequality and climate disruption will lead to unrest and mass migration of the poor from the countries in the tropical and arid and semi-arid belts most subject to negative climate change. Their destination is and, increasingly, will be Europe and North America, to the discomfort of those regions.

The current energy revolution, therefore, differs in important ways from its six predecessors. It is defined by a problem caused by success. It is partly anticipatory, i.e. it is about forestalling predicted disasters. Yet it is both immediate and accumulative. As I have noted, global CO_2 emissions from all sources equilibrate rapidly, but only disperse slowly over a period of >100 years (see Figure 15) [22][23][24][96][97]. Consequently, the impacts are experienced globally, irrespective of the geographic source of the gases. Resolving the threat involves both intergenerational and transcontinental and societal equity. What happens in London or Los Angeles impacts on Niamey and Kiribati. We are undermining the future, a future that depends on our total collective emissions, past, present and future, and the coherence of our global responses.

There is no evidence of a current or imminent shortage of fossil fuels. Despite forecasts of peak oil and peak cheap oil, these resources have proved to be so abundant and our technological ability to extract fossil fuels so resilient, that a convenient solution because of depletion is not on offer. Humanity needs to practise uncharacteristic self-denial. We must desist from burning and releasing the CO_2 from about two thirds of all the known hydrocarbon reserves to avoid catastrophe, an act of unprecedented collective international will and self-denial.

While there is some political consensus that concerted action is required, there is little agreement how the emission cuts should be implemented nor on whom, proportionally, these cuts should fall. Should the larger, i.e. the richer, emitters take a larger cut?

Should historic emissions be considered? Should all carbon emissions be taxed? Should there be an agreed carbon or GHG allocation to each human or to each country? How might it be monitored? How should we weigh the relative merits of emissions from the food chain with those from, say, international leisure travel? The current USA government policy is denial.

A number of factors make these challenges especially problematical and recalcitrant. The Industrial Revolution, as well as improving the standard of living of many, has placed enormous wealth in the hands of a few individuals, companies and countries, many enriched by hydrocarbon energy. They remain major players on the world's stock markets and fossil fuel reserves are their assets; indeed, they are still investing in exploration. It is scarcely surprising that companies and individuals use their resources to muddle the evidence and fund counter claims [98]. Their assets, wealth and status are at stake. From their commercial perspective, virtually all the negative impacts of fossil fuel use (pollution, health hazards, as well as some climate change extremes) are off-balance sheet externalities. An International Monetary Fund paper assessed these 'subsidies' (a mixture of direct and indirect subsidies and 'off-balance sheet', negative costs to society, e.g. premature deaths from pollution) as amounting to more than $5.4 trillion a year [94]. However, the successes of free-market capitalism have inspired a cadre of ideologues wedded to the infallibility of the free markets and the merits of individual enterprise and speculation. These two interest groups have combined to mount an effective campaign of disinformation. These 'merchants of doubt' [98] have slowed the responses in many countries, especially in the USA, Australia and the UK, where free-market ideology is strongest. As a result, several decades have been wasted (see Figure 14).

Fortunately, other entrepreneurs have seen a major business opportunity and are investing in low-carbon energy production and storage systems, and low-carbon technologies. These are developing rapidly, and their costs falling. These, one must hope, are the seeds of the seventh revolution!

No single dominant 'low-carbon' energy technology has yet emerged to power the seventh revolution. Indeed, the solution may differ from region to region depending on local resources and specific energy demands. It can be asserted with confidence that very low CO_2 electricity will be part of that solution and its generation from renewable energy sources, especially solar, wind and hydro, is growing rapidly. In appropriate locations, they are now economically competitive but, because of their intermittency, require complementary storage systems such as pump-storage schemes and batteries. These latter are also developing quickly and becoming cheaper. All renewable resources unfortunately have downsides as they are based on dispersed relatively low-energy flux sources. Some require specific chemical elements, e.g. rare earths and a number have often some negative impacts on habitats or landscape. There are no free meals and, ironically, we are revisiting and reinventing technologies such as wind and hydropower, first developed towards the end of the agrarian era.

The possible civil contribution of nuclear electricity generation remains unclear. This option was strongly favoured some fifty years ago and has been successfully deployed in France, but is proving problematical and increasingly expensive. A number of serious incidents, e.g. Sellafield, Three Mile Island, Chernobyl and Fukushima, have increased safety fears and contributed to higher design and building costs. Furthermore, the long-term problem of the +1,000 year storage of the radioactive waste from the current technology remains unsolved. Meanwhile, the military and terrorist threats from the explosive energy and radioactive fallout from nuclear fusion and fission remain as potent as ever, perhaps greater.

Carbon capture and storage (CCS), by sequestering the emitted CO_2 in some inert geological form, could permit fossil hydrocarbon use to continue, but the technology is unproven on the massive scale required (i.e. the absorption and safe storage of up to 35 G tonnes of CO_2 per annum). It is certain to add costs to the global energy budget and itself will require significant energy input. Much is being made of 'bio-energy with carbon capture and storage' (BECCS), as this method would draw down atmospheric CO_2 while producing energy and sequestering the CO_2, but, again, issues of practicality, scale and cost are unresolved. The area required for the biomass/forest growth might well compete with that required for food crops as the population exceeds 8 billion and climate change bites.

As noted, food production and land-use change account for about 25% of global GHG emissions, even excluding emissions from the production of fertilisers and other agrochemicals and the food chain. The official FAO figure must not blind us to the issue that, even if CO_2 emission from the energy chain falls dramatically (say 80% to 90%), food-chain emission will remain problematical. The main issue is not food miles but the emissions of CH_4 from ruminants (from meat and dairy production) and waterlogged soils, e.g. rice paddies, and N_2O released by soil microbes during nitrogen fertilisation, itself a necessity for high crop yields. The implications of the food preferences of the affluent, the food deficiencies of the poor, the cultural allegiances and religious practices found in many countries, e.g. certain animals being sacred or forbidden, and a human population growing at ~75 million per annum, means that the global outlook is at best highly problematical [100].

In some denier circles much is made of plant-growth promotion by additional atmospheric CO_2. This is a real effect under certain physiological photosynthetic regimes, e.g. in temperate C-3 plants, including tomato, wheat and rice. In these species, the 'perversity' of Rubisco, sometimes acting as a CO_2-emitting oxygenase instead of catalysing CO_2 fixation, is an issue. But the effect is minimal in species adapted to warmer and drier climates, e.g. C-4 plants such as maize and sorghum, or those exhibiting crassulacean acid metabolism (CAM) such as pineapples and cacti [100]. The weight of serious studies indicates clearly that, when all aspects are taken into consideration, the net effect of elevated atmospheric CO_2 and consequential climate change and warming will be lower yields and poorer food quality. Simply moving production north (or south) is implausible in terms of soil types and physical barriers. Also, even if feasible, it would itself result in the release of extra CO_2 through the oxidation of accumulated soil organic matter and tundra peats.

In this section I have focused almost exclusively on energy, but other human impacts are found throughout the bio-, hydro- and geospheres. These are so great that many consider that we have left the Holocene and have entered the Anthropocene age and are embarking on the sixth great extinction of species [101].

CHAPTER XI

On Human Behaviour and Our Social and Physical Constructs

Even the most superficial analysis of the seventh revolution shows that responding to its challenges extends well beyond the technical issues. In the late 1940s the United States designed and funded the ambitious postwar Marshall Plan to rebuild the economies of western European countries after the disaster of the Second World War and to escape the embrace of Stalin. In 1961 President Kennedy committed to putting a man on the Moon within the decade. In contrast, the political will and economic commitment to combat global warning has been tepid. It is worth recalling that the first major report on the subject was to the American President Lyndon Baines Johnson in 1965 [102].

To help understand this inaction and to find ways to avoid disaster, we must extend our analysis to a consideration of human aspects of the homeostatic hierarchy I alluded to earlier. As Antonio Damasio [20] suggested, and I have sought to elaborate, complex mechanisms, some largely automatic, have emerged to maximise 'wellbeing' of individuals at a personal and societal level. These must have evolved as our hominid ancestors evolved. These processes,

I suggest, still inform our behavioural and ethical choices. On these inherited behaviour traits, human communities have built ever more elaborate mechanisms to organise and perpetuate their communities and to manage and stabilise complexity.

Earlier, I sought to distinguish these innate behavioural mechanisms from the human constructs, such as laws and regulations, although, I grant, the two must be interconnected. I instanced the advent of cars and trains and the challenges and opportunities arising for them. I also recognised that opportunities for deception are offered by our technologies. Following Damasio, I suggest that the roles of our emotions and feelings in mediating and regulating our instinctive behaviours, so seeking to maximise 'wellbeing' and societal accord, are the human equivalents of cellular and whole-organism physiological homeostasis. In large measure, these largely automatic responses determine whether we, as humans, can co-exist in contentment or are primed for conflict. Given the complexity and ambiguity of human nature, we must recognise both our ruthless, competitive spirit and our capacity for selfless co-operation, and that we are subject to our social emotions of shame, contempt, sympathy and awe/gratitude (see Table 3).

A complementary and illuminating approach to the study of human behaviour has been led by Daniel Kahneman, Amos Tversky and Richard Thaler amongst others, and is summarised in Kahneman's bestselling *Thinking, Fast and Slow* [71]. Their work is, I believe, highly pertinent to the challenges and outcomes of the seventh revolution.

According to Kahneman and others, humans display two discrete mechanisms when responding to external events. Their evidence is compelling and derived from physiological (e.g. involuntary physical responses) and neurological (brain scanning) evidence and supported by many experiments on human behaviour.

We employ two 'thinking systems'; one, which Kahneman refers to as System 1, is rapid and intuitive and closely related to the Damasio's automated responses. It has discrete traits, biases and heuristics and determines most of human behaviour. The more analytical System 2, with which we are often loath to engage, is slower, more deliberative, demanding time and effort. The latter involves concentration and hard mental work and is not automatic. Generally, we prefer to avoid System 2, being, in this regard, lazy, as described by Morris [70]. In any case, System 2 deliberations are heavily influenced by our intuitive System 1 responses. At times we use System 2 not to make a dispassionate analysis of the choices facing us but to forge ready-made justifications for what our intuitive System 1 has already 'decided'. System 1 is itself subject to systematic errors, biases and manipulation, both by external stimuli and by our own memories and embedded experiences. In much the same way our feelings were the sum of our current, past and anticipatory mental processes in Damasio's analysis. Kahneman describes a number of our heuristics (simple but efficient mental shortcuts) and biases that influence, indeed often determine, our intuitive responses and choices. It seems reasonable to suggest that these have arisen to cope with living in small family or tribal groups, in an uncertain, often fraught, world for many hundreds of thousands of years. As hunter-gatherers in a hostile world, these two behavioural responses to externalities and to decision making must have been invaluable.

We tend to draw the most optimistic conclusions from the limited range of information available to us. Kahneman speaks of 'WYSIATI', i.e. 'what you see is all there is' and of our blindness to other possibilities and to information outside our normal range and perceptions. We tend, for example, to have very poor appreciation of statistical risk. We are very prone to a 'confirmation bias', i.e. we gravitate to and readily accept information that confirms our pre-existing views – an effect found every day in our choice of the daily papers or news channels. Disturbingly, this phenomenon is now being exploited by search engines such as Google as well as newspapers and politicians. As we browse, websites are preselected and prioritised by the search engine so as to conform to our existing biases, blinding us to the facts and opinions that we may find uncongenial. Our System 1 responses are strongly influenced by recent events, be they positive and cheering or negative and depressing. So, in terms of climate change, a sudden heatwave will increase 'belief' and a cold snap exacerbate 'doubt'.

The evidence, unsurprisingly perhaps, points to the influence of public opinion and pressure on our behaviour. We humans are occasionally dishonest and hypocritical and are as capable of self-deception and self-justification as of seeking to deceive others [103]. Behaviour studies show our honesty to be limited. It diminishes if we feel that we enjoy impunity or immunity and have the power or the opportunity to escape the consequences of our actions. The least corrupt countries, and incidentally those with highest well-being ratings, are those with open, well-informed, well-educated citizens (not subjects!), while the most corrupt are broken, lawless, often authoritarian regimes. In the latter, bribery and corruption are ingrained, so individuals, even if removed to a more open society, remain more likely to deceive and accept corruption as part of life. There is a strong link between wealth and power and the enjoyment of impunity. This results in individual greed at the expense of society, as revealed by the Panama papers [104] and in pervasive tax havens. In the context of the seventh revolution, many of those with 'wealth impunity' have the heaviest carbon footprints. Some are the direct recipients of coal, gas or oil wealth. In turn they have the most to lose from an equitable resolution, perhaps indeed any solution, of the seventh revolution. Their status and affluence appear at stake and they have the ability to dissemble, promote misinformation and exploit the social media to protect their self-interest and, maybe, in their own perceptions to buy themselves out of any trouble.

One of the results of these psychological studies has been the development of behavioural economics [71]. It shows that real-life choices are made by comparison to a reference point generated by the chooser. Thus, a gain of £100 from £900 to £1,000 is not perceived as equivalent to a gain from £100 to £200. They are, of course, equivalent in monetary value, and in classical economic theory our action should be governed by this simple maths. This, however, is not how we assess the risks and gains according to Prospect theory. It shows that we view the gains very differently to losses and are much more averse to losses than we are positive about prospective gains.

Further aspects of behavioural studies that are highly relevant to our reactions to the challenge of climate change are summarised by Jonathan Haidt, a moral psychologist [105]. He reinforces Kahneman's distinction between our rapid automatic intuitions and our slow strategic usually reluctant reasoning. He likens the former to an elephant and the latter to its rider. The riders' main purpose is to serve the elephant and occasionally provide guidance. However, in applying this reasoning to our moral choices and value judgements, Haidt adds some further insights. In classical Platonic philosophy, rationality is seen as the foundation of our choices. In contrast, David Hume and others have stressed that we are primarily led by our passions and often neglect reason. Haidt also draws on the work of Emil Durkheim [106] who saw humans, not as atomised individuals, but as embedded in their specific communities. Human wellbeing, Durkheim argued, depends on the strong and restraining relationships imposed by that society. Consequently, Haidt proposes that we must recognise that humans are often led by their passions, are intensely tribal or groupish, with a strong loyalty to their own group, and derive great satisfaction from being members of their 'hive'. We are indeed selfish individualistic primates but with a longing to be part of a group: a specific group, be it ethnic, national, religious, class or a sports team with which we can identify. We are hard-wired to enjoy feeling ourselves part of something larger, sometimes more noble, than ourselves. 'Morality binds, morality blinds' (page 28) [105]. Critically, there is no evidence that we normally relate to or identify with the whole of humanity. Our allegiance is to our 'group' or 'tribe' and is often exclusive. We, all too often, despise outsiders who may, or may appear, to threaten our personal or group identity and even worse our affluence – our entitlement.

Haidt developed a matrix by which he suggests we assess our moral 'choices' and value judgements, or, as he expresses it, define our 'moral foundations'. These are the characteristics, meritorious or otherwise, of his six paired responses: care/harm; liberty/oppression; fairness/cheating; loyalty/betrayal; authority/subversion; sanctity/degradation. He suggests that, if people are to be convinced of the righteousness of certain actions and, in our democratic system, vote for them and/or apply them in their own lives, most of the first named components of the above paired matrices must be engaged. In the case of climate change, it implies that the idea of equal shares globally will have little appeal. Even simple appeals to 'care' or 'fairness' will not broadly succeed if other 'moral foundations' are disregarded. Similarly, an appeal to the 'rights of the unborn generations' will carry little weight if it appears counter to other moral judgements. It may be that the recommended actions are perceived as oppressive or promoting cheating or calling for a seemingly disproportionate personal or group loss. The fear of betrayal by those outside the tribe and a capacity for subversion could be powerful counterforces.

More positively, one aspect highlighted by Haidt and others is the power of feeling part of a great project. As discussed in relation to the agricultural revolution, humans have an organisational, co-operative capacity to exploit their own kin and other animals and commit to undertaking vast enterprises, e.g. building great temples or irrigation schemes. These often appear to be undertaken exclusively on behalf of elites, but there is a growing recognition that they have contributed to the social and moral fabric of these societies. Joint enterprises can lead to the sanctifying of a place or a project and to conferring of sense of 'awe', so contributing to group loyalty and cohesion. Great joint projects, including wars, while reinforcing authority, have deep communal, psychologically cohesive impacts.

The aspects of human behaviour revealed above emphasise the difficulties we face in responding to anthropogenic global warming and I will return to this issue later. But here I must turn to one of the most pervasive of our human constructs – our current free-market economic system and how it influences our values and our responses to the seventh revolution.

I have noted that the beginnings of capitalism preceded the Industrial Revolution. Nevertheless, the availability of plentiful cheap controllable energy not only catalysed the global growth of capitalism but totally altered the balance of power between nations and continents. The concentrated power of firearms and gunpowder predate Watt's engine, but the capability of their mass production and their rapid deployment around the globe was a game changer. This is illustrated by Ian Morris's question [70]: why did the English dominate the Chinese in the late nineteenth century not vice versa?

The great growth of nineteenth-century capitalism was given an intellectual boost by the ideas that emerged from Darwinian natural selection and the competitive success of some species. Not only was 'nature red in tooth and claw' but Darwinian competition was seen as desirable, purifying and just in human society. Certain races and groups – obviously primarily white, Protestant males – were seen as the deserving winners in this competition for resources and wealth. This 'justified' their dominion over nature and over the 'lesser breeds' of humanity and, conventionally, women. In this Social Darwinism, a dangerous cocktail of biology and capitalism was created. Eugenics was one outcome, colonialism another. As Snyder [107] has pointed out, there is a direct link to Nazi ideology, to *lebensraum* and to the Holocaust.

Despite many horrendous events, the emergence of capitalism, science technology and, to an extent, democracy, has heralded an unprecedented growth in the global human population and in its total accumulated wealth as well as precipitating climate change. It is important, in my judgement, to seek to understand the processes, institutions and theories underpinning this, and how they are likely to influence our responses to the seventh revolution in both an energetic and homeostatic perspective.

While my background in economics is limited, I find Galbraith's classical analysis [108] compelling. He postulated over sixty years ago that continued affluence, when basic human needs are satisfied, depends on an implicit, largely unspoken, economic 'bargain' between three main parties: general public, industrialists and businessmen, and politicians. Each of these parties has a rational self-interest in adhering to this bargain (see Figure 15). We, the public, wish to retain and, if at all possible, increase our affluence. This depends heavily on the availability of jobs, preferably well paid, so as to maintain individuals and their families and to protect their buying power and social status. Jobs require flourishing businesses derived from the provision of goods and services and a competent, solvent government. The range and quantity of these goods and services must be expanded continuously through private and public innovation and investment; the additional demand can then feed through to new employment and good pay. These innovations might range from new information technology gadgets or new cars, to providing care to an ageing population or new drugs. The new affluence itself breeds new 'wants': holidays in the sun, holiday homes in Abersoch or the Algarve, gourmet foods, fine wines and the latest fashions.

Politicians realising 'it's the economy, stupid' must oil the wheels of this 'bargain' if they are to be (re)-elected by a relatively contented populace. Industrialists and businessmen and businesswomen have a strong vested interest in ensuring the continual turning of this cycle so that their businesses may prosper. To help sustain this bargain, multibillion industries such as advertising and product promotion, as well as novel designs, have grown up to convince the public of its urgent need for both traditional and innovative goods and services (see Figure 16) [109]. It has spawned brand loyalty, celebrity endorsement and a range of more subtle psychological strategies to stoke and stimulate our desires, now influenced heavily by the insights of the behavioural psychology I discussed earlier. If demand is waning or weak, it must be created!

Entirely reasonably, business has also lubricated the wheel by devices such as hire-purchase, easy credit, tempting special offers, all supplemented by ubiquitous credit and debit cards and loans. In many and subtle ways, people are encouraged to take on debt, most dramatically to buy and fit out a home, so fulfilling the dream of a self-reliant citizen in a property-owning democracy. Companies must themselves borrow to compete and to invest in new exciting

Figure 15: The 'Galbraithian bargain'

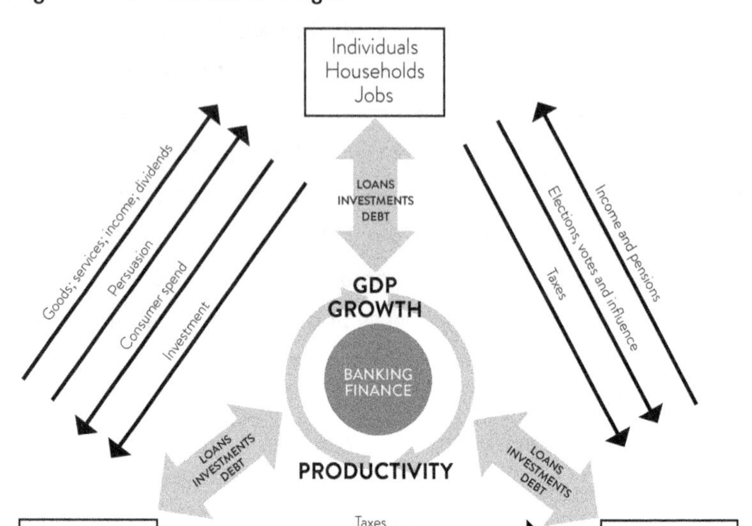

products, services or new outlets. Governments, in turn, must borrow to play their part in keeping the economic wheels turning by investing in education, infrastructure and dampening the business cycles so as to reduce social discontent and seek to ensure their own re-election and political survival (sometimes propping up failures to do so). Credit and debt lubricate the whole system, so giving huge power to financiers, who emerge as the puppet masters.

The cycle must turn continuously and its cycling is recorded as economic activity measured by gross domestic product or its derivatives. Critically, this must grow exponentially at an annual rate that varies widely from country to country. China has enjoyed a better than 7% annual rate, which implies a doubling of activity every decade. The UK dreams of an annual 3% growth rate but settles at one below 2%. Our economic model is nevertheless predicated on continuous exponential growth. The economy, as illustrated in Figure 15 [104], must spin like a top and grow like Topsy to retain its stability. As Galbraith foresaw [108], left to its own devices the 'bargain', to which all the parties have enthusiastically subscribed, ensures vested interests turn a blind eye to mounting debt. He saw the system as intrinsically unstable, leading inevitably to excessive, unsustainable debt, to overshoot and to the painful bursting of economic bubbles. He suggested that the natural tendency to overshoot was caused by over-optimism, aided and abetted by 'conventional wisdom' and by the human capacity for self-delusion delaying any appreciation of reality, with, of course, an added injection of greed and self-interested deceit. All these we now see as classical System 1 traits. 'Affluence', according to what I refer to as the 'Galbraithian bargain', requires a continually and indefinitely accelerating cycle of product innovation and consumer demand, supported by government and oiled by debt. Paradoxically, our economic success commits us to living on an ever-accelerating treadmill.

Galbraith emphasised the need for prudent regulation and careful economic management; this, I propose, is analogous to the homeostatic mechanisms discussed earlier. He also recognised a tendency to public squalor despite private plenty. He suggested that public finances would inevitably come under pressure as power and financial clout would accrue to a small elite and that the affluence of the few could reduce civic and collective pride and commitment, i.e. the human homeostatic mechanisms would be set to favour the elite.

After the crash of 2008, we can conclude unambiguously that our politicians and economists would have been well advised to take Galbraith's diagnosis much more seriously. However, central to his insight was the improbability of their doing so. Galbraith recognised that bankers, as intermediaries and the controllers of credit, wield enormous power, but he did not anticipate several factors. First, the scale of leverage and invention of the clever, risk-concealing packages that led to the 2008 crash. Secondly, the triumph, both in the USA and UK, of the cult of the infallibility of the unfettered market. Initially, under Reagan and Thatcher, and then Bush and Blair and Brown, the consequential rush to deregulation was quite contrary to Galbraith's insights. In the UK the 'big bang' in the City and in the USA the repeal of the Glass-Steagall Act by the Gramm-Leach-Bliley Act in 1999 helped ensure 'irrational exuberance' whose most influential cheerleader was, paradoxically, the chairman of the US Federal Reserve, Alan Greenspan himself. A few pertinent statistics: in 2010, global foreign currency transactions totalled $955 trillion; off-exchange trading in financial derivatives $601 trillion, although the traded volume of shares and bonds was $78 trillion, compared with the global gross domestic product of $63 trillion. These systems exploiting and leveraging vast sums of 'virtual' money have allowed hyper-rich individuals to emerge in the real world [93][94][108]).

Thirdly, especially in the UK, was the perception that government was reliant on the financial services sector for a significant part of their tax revenues and consequently on this sector to fund many public services. The sector became untouchable, despite Adair Turner's recognition, when chair of the Financial Services Authority in London, that much of their activity had no public benefit. Moreover, a recent report [110] suggests that the contribution of the financial sector to the UK Treasury is exaggerated and not confirmed by the audit trail in contrast to that from manufacturing which enjoys less support. Following the crash, direct government support to the financial sector was put at £298 billion and with loans and underwriting at £1.7 trillion. If these figures are correct, then either the City's public relations apparatus is quite exceptional or political self-interest is at play.

Fourthly, was the emergence of offshore havens that allow the rich and often the criminal as well as major transnational companies to minimise or entirely avoid scrutiny and regulation and their tax obligations in the countries where their business is undertaken. Shaxson [111] links this phenomenon, in large measure, to the successful attempt by the City of London to re-establish its global hegemony through the commercial exploitation of the residual British Crown Territories after the demise of empire. Such international financial gymnastics have been facilitated and empowered by a digitised, globalised world.

The fears expressed by Galbraith and Keynes [112] have been realised. The policy makers and politicians of successive governments in many countries – Ireland, Iceland, the UK and the USA, to name only four – have been seduced by their blind faith in the infallibility of the market and collusion with highly questionable banking practices.

Underpinning the global neoliberal economic consensus is an economic model based on *Homo economicus*, an atomised, self-centred, calculating creature whose evolution is damningly described by Raworth [95]. This view of humanity has informed the political economy of the world for several decades and permeated virtually all aspects of society with its presumption of the superiority of private gain and dismissal of non-monetary motivations. Behaviour studies have demonstrated our suggestibility and our tendency to react to the all-pervasive *Homo economicus* caricature by embracing his values.

Disturbingly, these trends are relevant to combating climate change and the seventh revolution in several ways – some already mentioned. The elite is able to divert huge sums, reckoned to be some trillions of dollars of assets, into offshore tax havens. This reduces the monies available for public goods, reduces the ability of governments to bridge the difficult change to a low-carbon economy, but increases the pressure to rev up the Galbraithian engine, as many have seen virtually no improvement in their living standards for decades. Keynes predicted [112] that we would be satiated by our affluence, but this has failed to materialise. Robert Skidelsky, Keynes's biographer, and his son Edward [113] have entered a plea that we urgently reassess our allegiance to material accumulation. They ask, 'How much is enough?' But there are obvious problems with their laudable response. The concentration of wealth, the export of work from the prime beneficiaries of the Industrial Revolution to the countries surfing the second and third waves, and automation have left a sizable minority in the rich countries in relative poverty and disillusioned with the Galbraithian bargain. The demand for food banks is not a mirage but we have been seduced by *Homo economicus* and embraced his values. The super abundance of the few has left a large minority disillusioned and with few benefits.

As described by Bernstein [114], trade has shaped and enriched the world. In the past, the main beneficiaries lived in the west. While the produce of the 'spice islands' catalysed early capitalism and the concept of the limited company in Amsterdam and London, the gain to the spice islanders was very modest indeed. Now the situation has been reversed. The erstwhile winners are easy prey to nativist propaganda seeking to blame others for their fate and all too happy to deny the reality of any threat such as global warming. The necessary reaction to the threat of climate change and the need to reduce the use of fossil fuels, such as coal burning, is painted as restraining the economy and the legitimate ambitions of *Homo economicus*.

In this situation it is also not surprising that many who were enjoined by politicians and the press and their own expectations to anticipate a continually rising standard of living have become disenchanted and rebellious. Ironically, their grievances and anger are fuelled by those committed to a free market or nativist ideology. Thus, despite an academic debate about zero-growth economies and the distortions implicit in measurement of gross national product, the central shibboleth is unchanged. Growth, frequently ignoring its environmental consequences, as exemplified by the pursuit of fracking in the UK, remains the politico-economic imperative.

However, a trend to new economic thinking appears to be gathering momentum. In her 'doughnut economics' [95], Raworth envisions the economy as being bounded by an ecological ceiling and bedded on a social foundation. It can flourish within a band constrained by the Earth's resources and by meeting the basic needs of all humans. Her model clearly also resonates with the theme of this book (see Figure 16) in showing energy fluxes as the drivers of life and the economy and as potentially conferring wellbeing

Figure 16: A model of the economy driven by energy but bounded by the Earth's resources

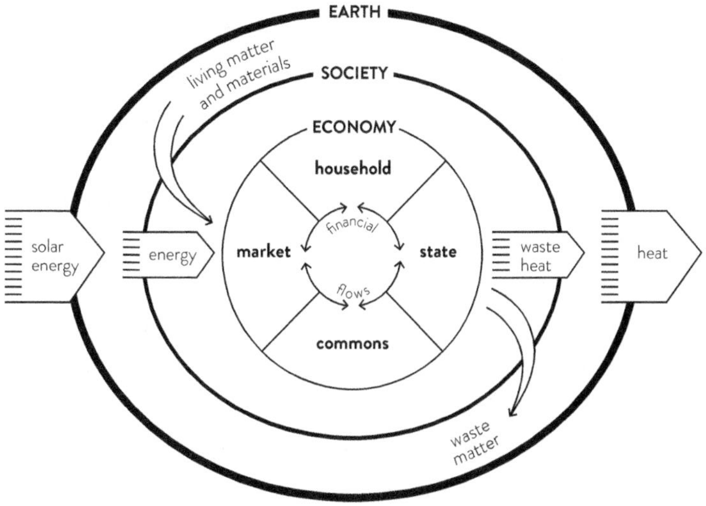

Reproduced with kind permission of Kate Raworth [95].

on humankind within the biosphere. She discusses the work of Robert Solow and his attempt to identify the factors responsible for the growth in the US economy in the first half of the twentieth century. Using Paul Samuelson's standard circular flow economic model, only some 13% of the observed growth could be accounted for. Much more recently Robert Ayres and Benjamin Warr [115] incorporated energy into the model, specifically the proportion of energy that can be harnessed to do useful work in addition to the classical elements of capital and labour. By doing so they were able to account for the vast majority of economic growth in several countries including the USA and provide another strand of evidence for one of the fundamental concepts underpinning this book (cf. the quotation from Ostwald in 1912 [6]). For a short discussion of some of the ideas relating energy to economic theory see note 16 and also the following references: [116][117][118].

Earlier, I alluded to the problem of quantifying the cost of homeostatic regulation in cells and whether it might pro rata increase with increasing complexity. While I am aware of no formal studies on the parallel issue in human society, it is possible a similar consideration will apply. There is no doubt that establishing social and economic homeostatic regulatory mechanisms must require an investment of energy and of human time, effort and ingenuity. It is not unreasonable to suppose that this might increase as ordered complexity itself increases. A case in point might be aviation and vehicular transport. More travel implies more and expensive runways and more sophisticated air traffic control et cetera. However, this has been, at least in part, counteracted by more improved, digitised control mechanisms. It appears that the capacity of computing and information technology to improve our regulatory systems has had two very different effects. On one hand, homeostatic regulation has become both more efficient and energetically cheaper (although fragile), and on the other, it has allowed the further speeding up of the pathway to complexity. Even so, that the regulation and stabilisation of complexity itself incurs an energy, material and social cost is a possible disincentive to effective intervention in the modern world. Certainly, modern regulation also offers opportunities for acquiring political power and wealth. Whose interests are embodied in the regulations? The rule of law is a vital homeostatic mechanism in human society, but perhaps often it has been appropriated to serve specific interests. It should of course be evenhanded and 'blind', but, given the facets of human nature discussed earlier, it is unsurprising that in many times and many places, it has been used to confirm and legitimise the power and status of elites and specific ideologies. In biological context, efficient homeostatic mechanisms confer a comparative advantage on an organism. In a socio-political context, it appears more likely that they confer a comparative advantage on a subset of humanity and can be too easily usurped by a few.

For some politicians and economists, Adam Smith's 'invisible hand' is suggested to provide an adequate, 'homeostatic' control over the workings of the market economy. But experience and economic history suggests, *pace* Galbraith and many others, that it does not. Clearly, given the current crisis, it lacks the capacity to drastically cut the GHG emissions or protect many social and environmental public goods. Externalities are too easily neglected, although they are capable of undermining our most vital, international public good – a healthy earth systems on which all life depends. It is one aspect of the Galbraithian prognosis writ large: the possibility of the private affluence of the few and a barely livable world for the many. Niall Ferguson [119] asserts that in modern society anything can be money; it is merely an immaterial confidence, sometimes confidence trick. Historically, money was proxy for the work and power derived from energy. But in some senses the creation of money has usurped energy as the great driver of human society, but the physical reality of energy demand remains (see, for example, Figure 16). Some calculate that their money can buy them a way out of climate change. Indeed, buying estates in New Zealand's South Island may bring a temporary reprieve but there will be no escape from global turmoil.

In summary, the evidence suggests that our intrinsic homeostatic behavioural responses and our elaborate social and economic constructs of the last 200 years are poorly adapted to the challenge of combating climate change. The former, while well geared to the lives of early humans and our predecessors for over 2 million years, are ill suited to solving an issue requiring foresight, an analysis of long-term risk, a global response and possibly self-denial. Any real answer must arise from our ingenuity and brain power, from a serious engagement with System 2, while being alert to our System 1 biases and heuristics. Our current socio-economic system, prioritising entrepreneurial self-interest, contributes to the dilemma in the myriad ways as previously outlined. It appears that our successful global economy, and its attendant models and assumptions, have become a likely cause of our undoing and fundamental to the problem. Paradoxically, given its power and expertise, it could, if fully engaged, contribute to the solution.

CHAPTER XII

Denouement?

This volume reaches back over 4 billion years to seek to chart the influence of energy on this planet and how the attendant work must be regulated and harmonised to ensure sustainability and a degree of wellbeing.

As I have noted, the time intervals between the energy revolutions have reduced dramatically and the speed at which they have rolled out across the planet decreased from billions to hundreds of millions, to tens of thousands, to 200 years. The work done per unit time (power) has increased, so accelerating the speed of change. The data in Figure 15 now show we have only two, at most three, decades to respond to the challenges posed by our current energy and economic regime. The great intertwined questions are: Will we rise to this challenge? What are the implications of the choices before us? Will the seventh revolution proceed smoothly? Will humankind move into a glorious new world? Or Will society go into decline, showing that our technical powers far exceed our wisdom?

In relation to energy, mankind appears to be riding a tiger. Without the two last revolutions, in all likelihood the genius of a Buddha or a Plato or indeed Newton or Einstein would have been unrecorded, confined to a small tribal clearing and forgotten in a few generations. Our understanding of this Universe would have been expressed in sagas and myths transmitted by word of mouth. Science would not have developed and the great works of human arts largely unseen and unheard. We would lack modern medicine and comfortable hygienic homes but also the capacity to alter the global climate and pollute our oceans and atmosphere.

To return to Nick Lane's question about the eukaryotic achievement – it has resulted in wondrous and beautifully complex organisms. But a price has had to be paid. The 'human advantage' has also disadvantaged many species other than those with whom we have made a Faustian bargain of their being husbanded in exchange for a short life. Some 12,000 years ago, humanity, in exchange for the social stimulation and variety of urban life and possibly more reliable but less nutritious food, started to forego the freedom of the hunter-gatherer. The behaviour studies discussed earlier have underlined the problems we now face because of the characteristic of our emotional and System 1 responses. Our intuitive emotions and reactions have evolved to achieve the most appropriate responses in small bands of hunter-gatherers over hundreds of thousands of years. Anthropogenic climate change, a consequence of our affluence and technical prowess, poses a completely new type of problem, one based on a threat to the global commons. It is statistical, anticipatory and likely contains 'unknown unknowns' as well as 'known unknowns'. Nevertheless, a reasoned, comprehensive, just and international response, fully engaging System 2, is required.

Given this background it is not apparent how our dilemma will be resolved. A number of scenarios can be envisaged:

1. Pessimistically, humanity will fail to respond in the required timescale and in sufficient depth. A major rise in mean global

temperature of >3°C degrees and other global changes will precipitate catastrophic consequences. The intricate global pattern of social and economic interdependencies will decay, quite possibly collapse. This would in all probability result in the death of millions and misery for billions. On a 100-year timescale, many of the human achievements since the agricultural revolution would be substantially reversed. Every individual, group and nation would be forced to fend and fight for themselves, in much the same way as the subsistence tribes described by Jared Diamond [61] protect their own precious resources. However, they would do so in a world of huge technological power. We could be faced with the frightening scenario of a resource-poor and population-rich, nuclear-empowered world where war would be endemic. The impacts on the cryosphere and on sea level would be profound with the seasonal, perhaps year-round, disappearance of the polar sea ice, the melting of glaciers and ice caps, and the loss of all-year water supplies to great metropolises and the inundation of coastal cities and communities. Whatever the physical environmental threats, there can be little doubt that the greatest dangers would arise from human behaviour. In *Collapse* [120], Jared Diamond has recounted how individual, localised societies have been overwhelmed by a complex admixture of factors. In this scenario this may be the fate of many communities and countries around the world.

2. Alternatively and optimistically, the common purpose and collective will, transiently apparent in Paris 2015, may still catalyse rapid action at a local, governmental and international level. Equally important, technological innovations driven by individuals, by growing social demands and by progressive government policies and binding international agreements will roll out

low-carbon energy supplies. These efforts will be reinforced by a serious commitment to energy saving. In this scenario, new technologies, humanity's collective commitment and inspired political leadership will solve the problem and seamlessly allow the current socio-political model of globalised capitalism, with minimal change, to triumph in a new low-carbon, quite possibly a lower energy world.

3. An alternative to this second scenario insists that technological and political 'fixes' will not suffice and that major changes in our lifestyles and economic and social priorities will be required. These will be even more challenging than the technical changes envisioned above and must include significant changes in our diets and lifestyle. Regrettably, it is difficult to envisage any voluntary pathway that will lead to such a sea change, given our political, economic and energy regimes and the human nature.
4. Human society may react grudgingly and slowly, partly because of the doubts sown by climate change deniers [98], partly because of our greed and short-sightedness, and partly because of the tragedy of the commons [121] and the tyranny of the present. As the seriousness of the climatic challenges and impacts become more and more obvious, the global community may panic and adopt a global energy 'marshall plan'. It may also resort to geo-engineering to try and draw excess CO_2 out of the atmosphere. Such actions must carry high risks and their outcome is entirely unpredictable. The inertia built into earth systems would still ensure that the mean sea level would continue to rise for centuries. There is, therefore, a very high possibility of serious unintended consequences and of precipitating major global unrest, e.g. if the Asian monsoon rains inadvertently were affected or major river flows, e.g. the Nile or the Yellow River, reduced as a result of such global

geo-engineering. In such a scenario, major conflicts are probable and the outcome may, in practice, be little better than Scenario 1.

5. Alternatively, current climate science could be demonstrated to be inaccurate and the impacts of the anthropogenic greenhouse gases proven to be much less than anticipated. The status quo and fossil fuel combustion would continue unabated. The seventh revolution would be delayed until fossil fuel reserves were exhausted. Indeed, the whole global-warming concept could be shown to be a huge plot to undermine capitalism, individual freedom and the American way of life. I note this, not because the evidence supports this contention, but because it is the one currently favoured by some powerful individuals, especially in the USA and the UK. It is the scenario favoured by free-market ideologues and some of 'my country first' protectionists. In science, the unexpected is always a possibility, but the odds of this scenario being valid are less than 1% and declining by the year. Whether it is sapient to take such a risk in the face of the evidence is the pertinent question. A supplementary scenario to this is the optimistic anticipation that 'something will turn up' and that some miracle cure will be found, e.g. volcanic cooling will miraculously balance out GHG warming: System 1-type optimism will be triumphant!

These scenarios are, in number of critical ways, oversimplified and even misleading.

The seventh revolution is being played out at a time when the digitised, electronic world of computers, virtual reality, automation, robotics and instant social media are reconfiguring not only work but many of humanity's traditional ways of assessing self-worth, wellbeing and sexuality. Historically, it seems probable that, from the time of *H. erectus*, 'human' wellbeing and self-worth have been intimately linked to work in, and for, close and extended families or small communities. Tribal allegiances were crucial to social status and to modest affluence. But this is changing in an atomised society.

Technology is miniaturising and growing ever more energy efficient. Whereas the great steam engines of the Victorian era were energy guzzlers, we now have sparse nano-machines and smart electronics. The historic relationship between energy and economic growth is changing, almost certainly for the better. But this increase in efficiency can readily be overwhelmed by an exponential increase in demand; computing and data storage are now important components of global energy demand.

The nature of work and community is also changing. Artificial intelligence and robotics are anticipated to take away many human jobs. Despite our affluence, the toll on humans in our highly competitive, even frantic, mobile consumerist world is becoming more and more apparent. Society is in many ways more fragile. Our electronic world can be destabilised by cyber-attacks or magnetic storms. Our just-in-time delivery systems are highly efficient but susceptible to interference. Few people know how to mend, still less build, all the gadgets their lives depend on, unlike our primitive ancestors. Many would be challenged to grow any of their own food. Many feel unable to cope and are suffering from 'affluenza' [122]. Others feel excluded, disillusioned and angry as the promised material rewards of capitalism have either passed them by or proven a disappointment, even a hoax. Some are stressed to the point of mental illness or drug addiction or suicide.

According to the historic precedents over 4 billion years, the seventh revolution would be anticipated to involve either the development of a transformational non-polluting energy source, and/or a step change in the energy economy through a leap in efficiency. Either would be anticipated to remove any current developmental ceiling and, by permitting more work to be achieved per unit time in all probability by more people, lead to a further acceleration in the overall conversion of energy into material, social structure and complexity. However, from the viewpoint of our common and historic humanity, this seems both improbable and highly undesirable. Undoubtedly, from the perspective of global biodiversity and mankind's environmental impacts, it would seem little short of disastrous. This implies that, in reality, even the optimistic Scenario 2 above may well solve very little.

I contend, therefore, that we are at a turning point. *H. sapiens'* continuing wellbeing now requires a complete reappraisal of our roles, aspirations and social constructs. In this sense the seventh revolution offers a great opportunity. Our relationships with our fellow humans and with all other life on this planet are threatening and are threatened. In this context, the exciting possibility of humans landing on Mars is a relative sideshow. Our futures are grounded on Earth. Can we therefore rethink and redefine what we collectively consider desirable and join John Gray [123] in seeing much of our current concept of 'progress' as a dangerous conceit? The issue is more ethical and social than economic or political or indeed environmental.

In his careful, restrained analysis of how some societies choose to fail while others survive [120], Jared Diamond suggests four main reasons for failure. The first is a failure to anticipate the arrival of the problem. This might be due, understandably, to a lack of experience of the issue or a long-forgotten experience and a false sense of security. Secondly, a society may fail to perceive a problem or its severity as it creeps up imperceptibly or is masked by natural fluctuations and by a lack of long-term communal memory or records.

Again, this is understandable in societies lacking writing and good historical records or scientific analysis. Thirdly, the problem may be perceived but its solution is at odds with what economists would term the 'rational behaviour' of an important interest group in that society, i.e. the priorities of *Homo economicus*. In such a case, a coherent response may be difficult to achieve and result in inaction. Finally, even when the problem is clear and a cure or cures are available, society may fail to take the necessary actions. They may be so wedded to a dogma, religion or psychological self-image that they fail, perhaps even prefer, not to react. As an example of the latter, Diamond alludes to the Christian Vikings of mediaeval Greenland who apparently refused to adopt Inuit ways and technologies, even to eat fish, when facing a worsening climate, 'preferring' to die out true to their traditions and in their self-absorption.

This deeply troubling analysis has so many resonances with the current reactions to global warming as to leave one overwhelmed. The impacts of climate change are masked by 'creep' and natural fluctuations and lie outside our recent experiences. The 'economic rational behaviour' of the oil, gas and coal companies and their political acolytes and well-funded think tanks, newspapers, websites and public relations companies is minimising, if not precluding, timely responses. The language of economic theory emphasises the ethical vacuum of their position, a position reinforced by the unswerving allegiance of a globalised elite to free-market dogma and a narrow interpretation of economic and social history.

This competitive system, as detailed in George A. Akerlof and Robert J. Shiller's *Phishing for Phools* [124], encourages deceit and dishonesty. One has only to consider the US subprime lending fiasco leading to the 2008 financial crash, or Volkswagen's lies about diesel emissions, or the manipulation of the LIBOR (London inter-bank offered rate) lending rates, or the misrepresentation of pharmaceutical trials, or the prevalence of drugs in sport, to conclude that phishing and cheating are endemic. They are not confined to, but are facilitated by, the worldwide web. It is reasonable to anticipate that, as/if the world seeks to mitigate GHG emissions and adapt to climate change by the seventh revolution, countries, companies and individuals will cheat. As Schumpeter pointed out: 'Economic progress, in capitalist society, means turmoil' [125]. But such turmoil has its dangers and may reduce the prospects of a non-catastrophic seventh revolution. Success requires global co-operation on an unprecedented scale, and must contain systems to minimise/prevent cheating (cf. Haidt's moral foundations [105]) and a degree of stability to allow people to cope with major societal change.

While this volume is focused on energy as the prime driver of change and human affluence, our attendant material demands are also impacting negatively on many vital aspects of earth systems, such as by tropical forest clearance, biodiversity loss, oceanic plastic pollution, atmospheric and nitrogen and phosphorus pollution, and soil erosion and degradation. These are creating both the Anthropocene era and a myriad of problems for human society and the biosphere.

One highly contentious issue is whether, under these circumstances, our democratic model can rise to the challenge. As has been amply demonstrated recently, alternative truths, post-truths and outrageous lies and distortions are commonplace and can be used to gull the electorate. As Daniel Kahneman has shown [71], our very nature is primed to fast, over-optimistic, barely considered reactions. We readily revert to the familiar and the comforting when facing uncertainly or an external threat. Plenty of politicians are happy to offer this false solace. Moreover, our political economy is based on the Galbraithian bargain, and on promises of continuous, unending exponential material growth. At best, such promises can only be fulfilled to 'winners' in a harsh economic competition. Necessarily, there are losers, in all likelihood a substantial majority of humanity. In our globalised society, where a few dozen own over 50% of all the world's wealth and resources *and* can hide their wealth in offshore tax havens [111], the rise of the losers, disaffected and disappointed seems inevitable. Surely a conflagrational scenario.

In this reading of the natural history of energy and the work and power so generated, the current energy revolution and the threat of global warming are harbingers of a wider global crisis and a clarion call for new ways of thinking and organising human society. This essay does not seek to offer any neat solutions but it does offer a specific long-term perspective. I contend that these threats are historic and deep-seated, perhaps the greatest challenge the 'wise ape' has had to face since cooking allowed neuronal numbers and brain capacity to grow so as to achieve modern cognitive and analytic functions. The crisis is born, paradoxically, of our dominance and success, of our abundant energy and cleverness. In the previous energy revolutions, existing ceilings to material and/or social complexity were shattered by new energy inputs. I doubt the seventh revolution will run a similar course.

Some obvious imperatives for action can readily be identified. Initially, we need technical fixes to develop efficient low-carbon energy systems to energise life while maintaining, with growing modification, the economy and seeking to live well on less. The introduction of carbon tax, globally, seems an essential prerequisite [126]. However, the difficulties implicit in such an aspiration should not be underestimated. Anthropogenic CO_2 emissions amount to about ~40 Gt annually and an IMF-supported report [99] suggests the hidden costs and subsidies related directly to these are about $5.4 trillion per year. This implies that fossil fuels cost the global community some $150 per tonne per capita per year. It would seem political insanity to raise a tax to combat these subsidies, but it is minimal condition for climate sanity for all fossil fuel subsidies to be removed. Secondly, we must secure food and water supplies for the growing world population, maybe 10 billion by mid-century. Achieving this, at low GHG emissions and monetary costs without destroying biodiversity and our soils, is again no small task. Thirdly, the question arises not only of energy and food supply but of its distribution, comparative fairness and respect for traditional lifestyles and beliefs. The latter is especially problematical given the large percentage of GHG emissions from ruminant animals. How do we deal with the ancient symbiotic dependence of the peoples of the world's highlands and semi-arid areas on their grazing animals? Can they be protected while meat and dairy production from intensive, but efficient feed-lots is heavily constrained, possibly discontinued? Fourthly, we must ensure that the digitised, globalised, wired-up world does not render humans workless and worthless. Fifthly, we must recognise that, while the immediate global socio-political hiatus may have little to do with energy per se, air pollution in major cities from fossil fuel burning is of increasing political concern and is killing tens of thousands. Paradoxically, solving the pollution crisis is likely to exacerbate and speed up global warming as the major pollutants contribute to climate dimming, but is nevertheless required.

In the medium term, climate change, energy supply and other social and economic issues are mutually interdependent and self-reinforcing problems that require an honest appraisal and serious response. Living on a small island off the north-west coast of the European mainland, likely to be spared the worst primary effects of global warming, does not offer any escape from global forces.

CHAPTER XIII

The Human Factor

ALBERT EINSTEIN: *We cannot solve our problems with the thinking we used when we created them.*
WALTER BAGEHOT: *The whole history of civilisation is strewn with creed and institutions which were invaluable at first, and deadly afterwards.*
ANTONIO GRAMSCI: *The pessimism of intellect and the optimism of the will.*
MARIO ANDRETTI: *If everything seems under control, you're not going fast enough.*

(Quoted in the *Independent* newspaper on 11 January 2017)

The conclusions of this analysis are not reassuring. On one hand our emissions of greenhouse gases are distorting the global balance of energy flows, so causing global warming and the addition and redistribution of heat energy in our oceans and continents.

Unfortunately, this is but one example of the unsustainability of the modern human society and its negative impacts on the Earth's ecology and regulatory systems. Nitrogen, plastic and chemical pollution, biodiversity and topsoil loss and land degradation and other impacts must be added to the list. Solving these problems in a world whose population is approaching 8 billion and whose economy is predicated on continual growth and resource mining is a daunting challenge.

I opened this monograph by referring to the tragedy of Aleppo, which, given my personal experience of one of our most ancient cities, resonates deeply with me. It illustrates all too dramatically how catastrophe can and does strike. However, the absence of serious social unrest in Cape Town and the collective effort to save water in face of dwindling reservoirs and avoid the taps running dry suggest a more optimistic scenario is possible. Regrettably, it appears that the non-catastrophic resolution of one of humanity's gravest problems, global warming, is made more difficult by nature of the homeostatic mechanisms that have historically modulated human behaviour. In our highly charged, digitised, atomised, global free market, I fear that the mechanisms, although well adapted to ensuring our survival as hunter-gatherers or as small farmers in small tight rural and even latterly in urban communities, seem inadequate, even redundant. Our automatic System 1-based tribal emotions and responses leave us ill-equipped to deal with global, 'statistical' issues that require cohesive forward planning by all mankind – and rapid action by our diverse and competing states and by our powerful market-driven transnational companies. We have made our Faustian pact with greed and individualism to drive growth and affluence, but it is increasingly clear that the consequences of this pact could be catastrophe. However, we are also capable of both intensive System 2-type analysis and actions, and of rising above the mundane and trivial, and of showing an appreciation of the greater good.

The relationship between our rationality and out instinctive responses and passions has exercised philosophers from Plato to Hume to today, and has been the inspiration for great sagas and much literature and art. These conflicts and tensions will remain. In appealing for urgent action, I am not suggesting that humankind will suddenly become 'perfectible' (cf. 123), but the nexus of energy, work and power combined with our technology and our current economic model can be changed if humans demand it.

Nevertheless, a hard-headed appraisal of our future prospects is extraordinarily difficult. It is a conceit characteristic of modern society, especially *Homo economicus*, that accelerating complexity can be equated to progress. This can only end in disaster. I believe it is also safe to say, unless a relatively just and fair but not necessarily equal allocation of global resources can be secured, conflict will only increase. Still, for all humans, hope and optimism are vital. It appears that much of humanity has been sold a false prospectus of a growing material affluence, hypothetically unlimited, unencumbered by any resource limitation and based on a socio-economic system that privileges a small elite. Politicians of a left-wing persuasion may seek to curtail such privilege, but most remain wedded to economic growth and tend to minimise the environmental risks.

Fortunately, our ability to access and husband this Earth's resources is not unchanging but can be enhanced by human ingenuity and efficiency, but requiring a circular not a throughput economic model. Our social and economic structures and moral and value systems are human constructs, and have been and can again be reinvented; even so our evolutionary heritage cannot be easily dismissed. Two points must be emphasised. The relationship 'energy – work/power- complexity' offers no moral compass. Energy, work and power has led to the slave sugar plantations of Jamaica, to the mass burials of the creators of the terracotta army, to Hiroshima and to the gates of Auschwitz. Until a few generations ago, physical labour defined the condition of humans, female and male alike. 'Arbeit macht frei' (work sets you free) was cynically arched over the gate of Auschwitz! Even before the Nazi death camps, work did not free but too often enslaved humans. Nevertheless, we are adapted to and often have defined ourselves through, our work.

Our understanding of the world around us and of our own behavioural psychology has grown dramatically over the last century. Furthermore, when humans decide to act, such as building the great monuments and temples of our ancestors or the US Moon programme, stunning achievements can occur over a relatively short space of time. Creating a Planet Earth that is a viable home for humans and all the other organisms with which we share her is surely the greatest project even contemplated. To start down this road, we need clear analyses of the issues and possible solutions. I hope that in placing the seventh revolution in the context of the six proceeding energy step changes over 4 billion years and a coherent regulatory homeostatic framework, I am contributing to this understanding.

The prime constraints appear to lie in our own natural history, as well as in our current politico-economic system; more in our personal values as in the hard science. Perhaps one can summarise the dilemma by asking: are we fated to mimic the tribal peoples of New Guinea and our ancient ancestors [61] who resorted to defending tribal assets? Must we retreat inwards in the face of a great global challenge? The trend toward societal fragmentation and the promotion of narrow self-interest is all too apparent in many, many countries and in recent elections and referenda. It bodes ill. However, such trends are not preordained or predestined and can be countered.

My hypothesis (see also note 17) suggests that, while over the millennia, energy has empowered work and, with increasing rapidity, resulted in material and social complexity, this conversion has reached its zenith. Otherwise, modern dominant *Homo sapiens*, the product of this long process, as well as the rest of the biosphere, will be the loser.

In the last paragraph of his last book [127] published shortly after his death in 1975, Arnold Toynbee, the grand old man of English historians, wrote:

> Will mankind murder Mother Earth or will he redeem her? He could murder her by misusing his increasing technological potency. Alternatively, he could redeem her by overcoming the suicidal, aggressive greed that, in all living creatures including man himself, has been the price of the Great Mother's gift of life. This is the enigmatic question which now confronts Man. (page 596)

The question remains unanswered.

Notes

1. Entropy: this is a rather 'slippery' concept. It is a statistical and additive property of a physical system and a measure of the kinetic energy contained in all the possible configurations of given molecules in a system at a particular temperature. It is expressed as energy per degree Kelvin, i.e. the temperature as related to absolute zero, usually per kilogram or per mole. Systems with a low entropy have high capacity for doing work. The concept arose initially from work on heat engines in the nineteenth century and the problems associated with understanding frictional, dissipative and irreversible heat loss. It was observed that there was inevitably some loss of heat energy during any cycles. However, later entropy was defined as a statistic mechanical property of a system, i.e. the sum of all the microscopic configurations of, say, the molecules in a gas compatible with its macro-properties, i.e. its volume, pressure and temperature. This led to an association of entropy both with order and information. The fewer the compatible states, the greater the order. As information is defined as the resolution of uncertainty, there are also implications for this field. The more we know about a system and the fewer the options, the less new information one is likely, statistically, to obtain about it.

   As mentioned, entropy is the thermal energy per unit temperate that is not available to do work. Systems with high entropy have a little capacity to do work. One corollary of the dissipative loss of heat is the long-term decline in useful entropy and the inevitable running down of the Universe to a state of quiescence – referred to as the 'heat death of the Universe'. On the other hand, living systems utilise the entropy from their surroundings to create an internal delineated space of low entropy and significantly greater order but at the expense of an inevitable dissipative loss of heat and an increase in the entropy of the external environment.

2. Homeostasis: the word is derived from two Greek words meaning 'like or resembling' and 'remain or the same' and was, according to Mario Giordano [8], first used in a biological context by Walter Bradford Cannon in *The Wisdom of the Body* in 1932, but the concept is far older. Claude Bernard, the father of animal physiology wrote of 'the stability of the internal environment being a condition for a free and independent life'. It may be argued that Spinoza, the seventeenth-century philosopher, anticipated the biology when he postulated: 'each thing [organism] as far as it can by its power strives to preserve its being' (Part Three of Ethics Proposition VI; in *Spinoza Selections*, ed. John Wild (New York: Charles Scribner's Sons, 1953.), p. 214.)

   The central biological concept is that mechanisms exist, indeed must exit, at all organisational levels, ranging from a single cell to humans and latterly their constructs, if the structural complexities arising for the work done by harnessing energy fluxes are to be stabilised. Generically, I suggest, homeostasis is the process by which structures and systems, generated by work achieved by the exploitation of energy fluxes, are, largely automatically, stabilised and sustained out of equilibrium with their environment. A very similar concept, which he calls informational metabolism, is explored by Kępiński (see [83]).

   Cell biology provides compelling evidence that the core cellular functions (see note 3) tend to be conserved around a set of defined biochemical and biophysical conditions. In turn the fundamental consistency of living organisms has allowed the genetic identity of the earliest prokaryotic and eukaryotic common ancestors to be deduced (see note 4). Nevertheless, natural selection has allowed limited deviation from these 'set-points', e.g. all cytoplasms are K^+-specific but the concentration may vary (more widely in prokaryotes), and the evolution of a huge range of additional capacities and cell differentiation and specialisation with this framework.

3. Translation and transcription: to simplify greatly, the genetic data encoded in DNA is first transcribed into a messenger RNA, which is then translated into the correct order of amino acids that fold, often after modification, into the resultant active protein. This takes place on biochemical nano-machines called 'ribosomes', which themselves contain another form of RNA, and whose structures have been highly conserved since the LUCA. These processes have specific ionic, energy

and catalytical requirements such as relatively high concentrations of potassium ions, magnesium ions, and ATP but are disrupted by modest concentrations of ions such as sodium commonly found extra-cellularly in seawater or the mammalian blood stream.

4. LUCA: the last universal common ancestor of all cells and consequently of all living organisms, pro- and eu-karyotic alike. The genetic make-up and characteristic of this cell are deduced by a statistical genetic back-casting technique that allows the genes most likely to have been present at the beginning and their activities to be reconstructed. In much the same way, the characteristics of LECA, the last eukaryotic common ancestor, has been deduced and the inheritance of mitochondria attributed to a bacterial and other cell components to an archaean ancestor.
5. Definition of life: The definition of life is not as straightforward as it might appear. All organisms are built up of cells which are capable of metabolism so that they can grow, react and replicate. All depend on importing material and energy e.g. sunlight or forms of chemical energy, from the external environment. Problems of definition arise in relation to viruses which can only replicate their genes/DNA by 'parasitising' on the cellular metabolic capabilities of other cells. Although they are replicating genetic entities, should they be considered alive? This situation can be viewed as an extreme case of co-dependence commonly observed in biology as virtually all organisms are dependent on the functioning of others e.g. for fixed or available nitrogen or for carbon or vitamins. However this is resolved there is little argument that, during part of their life cycle, maintaining homeostatically-regulated, metabolising cells, drawing on but out of equilibrium with their environment, is an essential feature of living entities.
6. Thermophoresis: this is a process by which over very small distances diffusive gradient can be established, which leads to the concentration of organic molecules in a particular location.
7. Microbial nomenclature: the ways by which bacteria can grow depend, primarily, on the sources of carbon, and the availability of reducing power and of utilisable energy. Thus, microbial metabolism can be arranged according to these three principles: (a) sources of carbon: **autotrophs** are able to fix gaseous CO_2 in the air but **heterotrophs** obtain their carbon from pre-existing organic compounds; (b) sources of reducing power: **lithotrophs** obtain reducing power from inorganic compounds while **organotrophs** tap into the potential of reduced organic compounds; (c) sources of energy: **chemotrophs** obtain energy from existing

external chemical compounds, while **phototrophs** rely on energy from light (photons).Typical examples are chemolithoautotrophs that obtain energy and reducing power from the oxidation of inorganic compounds and carbon from the fixation of carbon dioxide, e.g. nitrifying bacteria, sulphur-oxidizing bacteria, iron-oxidizing bacteria or photolithoautotrophs that obtain energy from light and carbon from the fixation of carbon dioxide, using reducing equivalents from inorganic compounds such as water, for example cyanobacteria.

8. Redox gradients: energised negatively charged electrons (e^-) can be donated by a source with a negative redox potential, and the energy of that electron can be harvested when the electron is accepted by an acceptor molecule with a lower or positive potential, i.e. e^- flows from negative to positive. As is illustrated in the text, if the donor is glucose being oxidised to CO_2 and the acceptor is oxygen (being reduced to water (H_2O)) within a redox gradient, the energy potential gain is nearly maximised. But there are many more possible donors and acceptors, many of which have been exploited in biology. In all cases, the electron flow is coupled to the creation of a proton (H^+) gradient, which is used to synthesise ATP.
9. Archaea and bacteria: although bacteria and archaea are morphologically indistinguishable, they differ in the chemistry of their cell walls and their external bounding membranes. They also have differences in their ribosomal machinery with that of the archaea most resembling that of eukaryotes. Archaea were first found growing in very extreme conditions of very high or very low pH, very high temperatures or salinity. At one time this 'extremeophil' trait was thought to characterise them, but over recent years, they have been found virtually everywhere, including in humans and the rumen of cattle.

   Interestingly, although an ATP synthase enzyme is found in the archaea, it does not possess the characteristic mechanism of the ATP synthases found in bacteria, mitochondria and chloroplasts, as outlined in the main text. Nevertheless, it is fair to emphasise that the archaea generate ATP utilising a proton gradient and an ATPase synthase is similar but not identical to an F-type proton H^+-ATPase of other organisms. So, the basic mechanism is retained although the protein complexes have diverged during evolution.

   The similarities and dissimilarities noted are compatible with eukaryotes arising from the absorption of a bacterial donor which become a mitochondrion, as discussed in chapter IV.

10. Multi-cellularity: prokaryotic cell chains. All complex multi-cellular biological structures are built of eukaryotic cells. However, as always in biology, there are ambiguities and possible exceptions. Some cells in nitrogen-fixing species in the prokaryotic genus Anabaena of Phylum Cyanobacteria can undergo limited differentiation under nitrogen-limiting conditions. Under such circumstances, cells called 'heterocysts' are found in the filamentous chain of the organism. These have different properties to the other cells in the chain. So, it may be argued that the organism is multi-cellular but all cells are in direct contact with the external medium and the differentiation is very limited.
11. Rubisco: with the evolution of land plants, they were dependent on the flow of water from the soil to their roots and up to their leaves to draw up both water and nutrients and to sustain the turgor (hydrostatic) pressure of the leaf cells. The turgor pressure is the pressure within a walled plant cell that is required for extensive growth and many other physiological functions. In such a context, the dual character of Ribulose bisphosphate carboxylase-oxygenase (Rubisco), i.e. its capacity to catalyse reactions with both gaseous CO_2 and O_2, has posed significant problems [43]. The control of water loss from the plant's leaves by the opening and closing of their small valve –like stomata – is critical, especially in the drier parts of the world. Closing these 'valves' to limit water loss also limits the movement of air carrying CO_2 into the interior of the leaf, so too decreasing the internal CO_2 concentration. This in turn promotes the oxygenase function that leads, de facto, to carbon dioxide loss and a lowering of the effective rate of photosynthesis.
12. Mitosis and meiosis: in mitosis a single cell, which contains two copies of all its genes, divides to create two daughter cells, which contain the same genetic information, i.e. DNA sequences, in the same number and arrangement of chromosomes as found in the original mother cell. The original mother cell and its daughters are termed 'diploid', i.e. carrying two copies of each gene. In meiosis, two cell divisions take place leading to four daughter cells, each with only one copy of the genes, i.e. a haploid cells. These haploid cells from male and female donors (or gametes) can then fuse to form a new diploid cell (a zygote). The incipient new organism consequently carries one gene set from both parents. During the process of meiosis, the crossover of portions of the chromosomes can take place ensuring genetic mixing, i.e. the new organism is not an exact copy of its parents. Consequent variation is one of the sources of the genetic selection lying at the heart of Darwinian natural selection.

13. Neanderthal brains: neanderthal brains have a volume of about 1,600 to 1,300 cm^3, with the male volume being recorded as a little larger than the female volume. The mean volume for *H. sapiens* brains is about 1,200 to 1,450 cm^3. There are suggestions that the cranial volumes of *H. sapiens* have been decreasing over the years. These observations suggest that volume per se is a poor indicator of braininess but also that we should not be dismissive of neanderthal abilities.
14. Renewable energy: historically humankind has relied on a limited spectrum of renewable energy resources – mainly annual or recent photosynthetic wood, fodder or food. In the Roman, Chinese, Persian and other empires from about 3,000 years ago, some use was made of wind and water power. Geothermal heat was used in baths on a small scale. Coal and tar were also used in some locations. It is worth emphasising (see Figure 6) that only at the beginning of the twentieth century did coal become more important as a global source of energy than biomass, and that even today energy from biomass is more quantitatively significant than nuclear power.
15. See *https://libquotes.com/matthew-boulton*. This quotation is now seen on £50 bank notes! See *http://news.bbc.co.uk/1/hi/scotland/glasgow_and_west/8075130.stm* (last accessed 9 April 2019).
16. Energy and economics: unlike classical economics, ecological and environmental economics treat the economy, both global and local, as embedded in the energy and material/chemical fluxes of the biosphere. These latter are recognised to depend on incoming solar radiation and the litho-, atmos- and hydro-spheres and that the Earth's resources are not infinite. The detailed views of the most important promoters of these concepts are worth consulting, e.g. Nicholas Georgescu-Roegen (*The Entropy Law and the Economic Process*) [116], Schumacher (*Small is Beautiful: Economics as if People Mattered*) [117], and Robert Costanza and Herman Daly et al. (*An Introduction to Ecological Economics*) [118]. Kate Raworth's model (see Figure 16) lies in this tradition, with an explicit recognition of the social imperative to limit impoverishment within a system of constrained resources. Such ideas are not entirely new and can be traced back to John Stuart Mill and Rousseau, but are more relevant than ever.
17. A summary of this hypothesis was published previously: see [128].

References

1. R. Gareth Wyn Jones et al., 'A Hypothesis on Cytoplasmic Osmoregulation', in E. Marré and O. Ciferri (eds), *Regulation of Cell Membrane Activities in Plants* (Amsterdam: North Holland, 1977), pp. 121–36.
2. R. G. Wyn Jones, 'Cytoplasmic Potassium Homeostasis: A Review of the Evidence and its Implications', in D. M. Oosterhuis and G. A Berkowitz (eds), *Frontiers in Potassium Nutrition* (Norcross: Potash and Phosphate Institute, 1999), pp. 13–22.
3. Roger Leigh and Gareth Wyn Jones, 'Cellular Compartmentation in Plant Nutrition: The Selective Cytoplasm and the Promiscuous Vacuole', *Advances in Plant Nutrition,* 2 (1986), 249–79, 249.
4. David J. C. Mackay, *Sustainable Energy – Without the Hot Air* (Cambridge: UIT, 2008).
5. Vaclav Smil, *Energy and Civilization: A History* (Boston: MIT Press, 2017).
6. Wilhelm Ostwald, *Der energetische Imperativ* (Leipzig: Academische Verlagsgesseiahaft, 1912): *https://archive.org/details/derenergetische00ostwgoog* (last accessed: 9 April 2019).
7. Howard T. Odum, *Environment, Power and Society* (New York: Wiley-Interscience, 1971), p. 43.
8. Mario Giordano, 'Homeostasis: an underestimated focal point of ecology and evolution', *Plant Science,* 211 (2013), 92–101.
9. John Maynard Smith and Eörs Szathmáry, *The Major Transitions in Evolution* (Oxford: Oxford University Press, 1997).
10. Tim Lenton and Andrew Watson, *Revolutions that Made the Earth* (Oxford: Oxford University Press, 2011).
11. James Lovelock, *The Vanishing Face of Gaia* (London: Penguin, 2010), and his *A Rough Ride to the Future* (London: Allen Lane, 2014).
12. Richard Dawkins, *The Selfish Gene* (London: Oxford Landmark Science, 1989), and his *The Ancestors Tale* (London: Phoenix, 2004).

13. Matt Ridley, *Genome: The Autobiography of Species in 23 Chapters* (London: Harper Perennial, 2000).
14. Steve Jones, *The Language of Genes* (London: Flamingo, 1993).
15. Freeman Dyson, *Origins of Life* (Cambridge: Cambridge University Press, 1993). See also R. Gareth Wyn Jones, 'Sylfeini Biocemegol Bywyd, yn Y Creu', *Y Gwyddonydd*, 16, 2/3 (1978), 104–12, and [27], [28] and [35].
16. Bruce Alberts, *The Molecular Biology of the Cell: Sixth Edition* (New York: Garland Science, 2014).
17. R. Gareth Wyn Jones, Colin J. Brady and J. Spiers, 'Ionic and Osmotic Relations in Plant Cells', in D. L. Laidman and R. G. Wyn Jones (eds), *Recent Advances in the Biochemistry of Cereals* (London: Academic Press, 1979), pp. 63–103. p. 63. See also Janet M. Wood, 'Bacterial Responses to Osmotic Challenges', *Journal of General Physiology*, 145, 5 (2015), 381–4.
18. T. S. Gibson, J. Spiers and Colin J. Brady, 'Salt tolerance in plants. II In vitro translation of mRNA from salt-tolerant and salt sensitive plants on wheat germ ribosomes. Responses to ions and compatible solutes', *Plant Cell Environment*, 7 (1984), 579–87.
19. Ilya Prigogine, *La thermodynamique de la Vie. La recherché* (1972). Quoted in Schoffeniels [28] below (pp. 101–2, p. 107). Also: *https://www.nobelprize.org/uploads/2018/06/prigogine-lecture.pdf* (last accessed: 9 April 2019).
20. Antonio Damasio, *Looking for Spinoza* (London: Vintage, 2004).
21. Steven Rose, *Life-lines: Biology Beyond Determinism* (New York: Oxford University Press, 1997).
22. Intergovernmental Panel on Climate Change, Fifth Assessment Synthesis Report (2016). See *https://www.ipcc.ch/assessment-report/ar6/* (last accessed: 9 April 2019).
23. John Houghton, *Global Warming: The Complete Briefing, Fifth Edition* (Cambridge: Cambridge University Press, 2015).
24. Kevin Anderson and G. Peters, 'The trouble with negative emissions', *Science*, 354, 3609 (2014), 182–3.
25. Edward O. Wilson, *Consilience: The Unity of Knowledge* (London: Little Brown and Company, 1998).
26. Alexandra Witze, 'Oldest traces of life on Earth may lurk in Canadian rocks', *Nature* (27 September 2017): *doi:10.1038/nature.2017.22685*.
27. Jacques Monod, *Chance and Necessity* (Glasgow: Fontana, 1974).
28. E. Schoffeniels, *Anti-chance* (Oxford: Pergamon, 1976).
29. Nick Lane and William F. Martin, 'The origin of membrane bioenergetics', *Cell*, 151 (2012), 1408–16.

30. Stanley Miller and Harold C. Urey, 'Organic compound synthesis on the primitive earth', *Science*, 130, 3370 (1959), 245–51.
31. A. Oparin and J. B. S. Haldane: *https://en.wikibooks.org/wiki/Structural_Biochemistry/The_Oparin-Haldane_Hypothesis* (last accessed: 9 April 2019).
32. J. Hunter Waite et al., 'Cassini finds molecular hydrogen in the Enceladus plume: evidence of hydrothermal processes', *Science*, 360, 6385 (2017), 155–9.
33. Thomas R. Cech, 'The RNA World in Context', *Cold Spring Harbor Perspectives in Biology*, 4, 70 (2012). See *https://cshperspectives.cshlp.org/content/early/2011/02/14/cshperspect.a006742.full.pdf+html* (last accessed: 9 April 2019).
34. Robert J. P. Williams, 'The Chemical Elements of Life', *Journal of the Chemical Society, Dalton Transactions* (1991), 539–46.
35. Nick Lane, *The Vital Question: Why is Life the Way it is?* (London: Profile Books, 2015). See also William F. Martin, 'Ancient Cells: Going back in Genes', *The Biologist*, 64, 2 (2017), 20–3.
36. Peter F. Mitchell, 'Coupling of phosphorylation to electron and hydrogen transfer by a Chemi-Osmotic type Mechanism', *Nature*, 191, 4784 (1961), 144–8.
37. Ira G. Wool, Y. L. Chen and A. Gluck, 'Mammalian ribosomes: the structure and the evolution of the Proteins', in J. W. B. Hershey et al. (eds), *Translation Control* (New York: Cold Spring Harbor, 1996), pp. 685–731.
38. George E. Fox, 'Origin and evolution of the ribosome', *Cold Spring Harbor Perspectives in Biology*, 2, 90 (2010). See *https://cshperspectives.cshlp.org/content/2/9/a003483.full* (last accessed: 9 April 2019).
39. Erwin Schrodinger, *What is Life? And Mind and Matter* (Cambridge: Cambridge University Press, 1967).
40. J. W. Lengeler, G. Drews and H. G. Schlegel (eds), *Biology of Prokaryotes* (New York: Wiley, 2009).
41. John A. Leigh, 'Nitrogen Fixation in Methanogens: The Archaeal Perspective', *Current Issues in Molecular Biology*, 2, 4 (2000), 125–31.
42. Lee Sweetlove, 'Number of Species on Earth Tagged at 8.7 million', *Nature* (23 August 2011). See *https://www.nature.com/news/2011/110823/full/news.2011.498.html* (last accessed: 9 April 2019).
43. Rudolf Amann and Ramon Rosselló-Móra, 'After all, only millions', *mBio*, 7, 4 (2018), 300999–16: *doi:10.1128/mBio.00999-16*.
44. Tuo Shi and Paul Falkowski, 'Genome evolution in Cyanobacteria', *PNAS US*, 105, 7 (2008), 2510–15.

45. Heinrich D. Holland, 'The oxygenation of the atmosphere and oceans', *Philosophical Transactions of the Royal Society*, 361 (2006), 903–15, *http://rstb.royalsocietypublishing.org/content/361/1470/903.full.pdf*. And *https://en.wikipedia.org/wiki/Geological_history_of_oxygen#/media/File:Oxygenation-atm-2.svg* (last accessed: 9 April 2019).
46. Nick Lane, *Oxygen: The Molecule that Made the World* (Oxford: Oxford Landmark Science, 2016).
47. Park S. Nobel, *Physicochemical and Environmental Plant Physiology* (Burlington: Academic Press, 1991).
48. S. Bengtson et al., 'Three dimensional preservation of cellular and subcellular structure suggest 1.6 billion-year old crown group red algae', *PLoS Biology*, 15, 3 (2017): e2000735.
49. Lynn Margulis et al., 'The Last Eukaryotic Universal Common ancestor; Acquisition of cytoskeletal motility form aerotolerant spirochetes in the Proterozoic Eon', *PNAS USA*, 103 (2006), 13080–5.
50. Tom Cavalier-Smith, 'Deep phylogeny, ancestral groups and the four ages of life', in *Personal Perspectives in the Life Sciences for the Royal Society's 350th Anniversary* (London: Royal Society Publications, 2010), pp. 111–32, p. 111.
51. Stephen Jay Gould, *Wonderful World: The Burgess Shale and the Nature of History* (New York: W. W. Norton, 1989).
52. S. Conway Morris, 'The Cambrian "Explosion": Slow-fuse or Mega-tonnage?', *PNAS*, 97, 7 (2000), 4426–9.
53. Christopher Field, M. J. Behrenfeld, J. T. Randerson and P. Falkowski, 'Primary Production of the Biosphere: Integrating Terrestrial and Oceanic Components', *Science*, 10, 5374 (1998), 237–40.
54. Philip Pullman, *The Amber Spyglass* (London: Scholastic, 2001).
55. Human lineage, *Encyclopædia Britannica:https://www.britannica.com/science/human-evolution/images-videos/media/275670/73009* (last accessed: 9 April 2019).
56. Chris Stringer, *The Origin of Our Species* (London: Allen Lane, 2011).
57. Jean-Jacques Hublin et al., 'New fossils from Jebel Irhoud, Morocco and the pan-African origin of Homo sapiens', *Nature*, 546 (8 June 2017), 289–92. See also *https://www.smithsonianmag.com/science-nature/earliest-humans-remains-outside-africa-just-discovered-israel-180967952/* (last accessed: 9 April 2019).
58. Suzana Herculano-Houzel, *The Human Advantage: A New Understanding of How our Brain became Remarkable* (Boston: MIT Press, 2016).
59. Richard Wrangham, *Catching Fire: How Cooking Made us Human* (London: Profile Books, 2009).

60. Walter Scheidel, *The Great Leveler: Violence and the History of Inequality* (Princetown: Princetown University Press, 2017).
61. Jared Diamond, *The World Until Yesterday: What we can Learn from Traditional Societies* (London: Penguin, 2013).
62. Daniel Everett, *How Language Began: The Story of Humanity's Greatest Invention* (London: Profile Books, 2017).
63. Noam Chomsky, *Aspects of the Theory of Syntax* (Boston: MIT Press, 1965).
64. Rebecca L. Cann, Mark Stoneking and Allan C. Wilson, 'Mitochondrial DNA and human evolution' *Nature*, 325, 6099 (1987), 31–6: 325031a0.
65. A. S. Wilkins, R .W. Wrangham and W. T. Fitch, 'The 'domestication syndrome' in mammals: a united explanation based on neural crest behaviour and genetics', *Genetics*, 198, 4 (2014), 1771. See *https://doi.org/10.1534/genetics.114.165423* (last accessed: 9 April 2019).
66. James C. Scott, *Against the Grain: A Deep History of the Earliest States* (New Haven: Yale University Press, 2017).
67. Yuval N. Harari, *Sapiens: A Brief History of Humankind* (London: Vintage, 2011).
68. Jared Diamond, *Guns, Germs and Steel* (London: Vintage, 1998).
69. Peter R. Shewry, 'Wheat', *Journal of Experimental Botany*, 60 (2009), 1537–53. And M. Feldman, 'Wheats', in J. Smart and N. W. Simmonds (eds), *Evolution of Crop Plants* (Harlow: Longman Scientific and Technical, 1995), pp. 185–96.
70. Ian Morris, *Why The West Rules For Now* (London: Profile Books, 2011).
71. Daniel Kahneman, *Thinking, Fast and Slow* (London: Penguin, 2011).
72. Hugh Thomas, *An Unfinished History of the World* (London: Pan, 1981).
73. David Landes, *The Unbound Prometheus* (Cambridge: Cambridge University Press, 1969).
74. See *https://www.ipcc.ch/publications_and_data/ar4/wg3/en/ch7s7-4-3-2.html* (last accessed: 9 April 2019).
75. See *https://www.e-education.psu.edu/earth104/node/1347* and *https://www.google.co.uk/search?q=graphs+of+global+population+and+energy+growth&tbm=isch&source=iu&ictx=1&fir=e_ufT8XY4CHATM%3A%2CyL4WTcYwXgBHwM%2C_&usg=AI4_-kQQF6VbN097LVYZcDyrasOhgCZbvQ&sa=X&ved=2ahUKEwj149W2wrbeAhUIGuwKHYkVBQUQ9QEwDHoECAAQCg#imgrc=U5cQ775IpjdaBM* (last accessed: 9 April 2019).
76. B. De Long, *Estimating World GDP* (2006): *http://delong.typepad.com/print/20061012_LRWGDP.pdf* (last accessed: 9 April 2019).
77. A. Madison, *Contours of the World Economy 1–2030 AD in Macroeconomic History* (Oxford: Oxford University Press, 2007).

78. See *https://ourfiniteworld.com/2012/03/12/world-energy-consumption-since-1820-in-charts/* (last accessed: 9 April 2019).
79. See *https://climate.nasa.gov/climate_resources/24/graphic-the-relentless-rise-of-carbon-dioxide/* (last accessed: 9 April 2019).
80. Christopher P. Kempes et al., 'Drivers of bacterial maintenance and minimal energy requirements', *Frontiers in Microbiology*, 8, 31 (2017), 1–10.
81. Richard Schiffman, 'Buzz Off', *New Scientist* (9 June 2017), 40–1.
82. Antonio Damasio, *The Strange Order of Things: Life Feeling and the Making of Cultures* (New York: Pantheon, 2018).
83. Jan Ceklarz, 'Revision of Antoni Kępiński's concept of information metabolism', *Psychiatria Polska*, 52, 1 (2018), 165–73.
84. Earl Cook, 'The Flow of Energy in an Industrial Society', *Scientific American*, 225, 3 (1971), 134–47.
85. William D. Hamilton, 'The moulding of senescence by natural selection', *Journal of Theoretical Biology*, 12 (1966), 12–45.
86. Steven Pinker, *The Better Angels of Our Nature: A History of Violence and Humanity* (London: Penguin, 2012).
87. Herbert Spencer, *The Principles of Biology* (University of the Pacific Press, 1863). See also, for critical analysis of this view, Raymond Williams, 'Social Darwinism', in *Cultural and Materialism* (London: Verso Press, 2005), pp. 86–102.
88. Marc Imhoff and Lahouari Bounoua, 'Exploring global patterns of net primary production carbon supply and demand using satellite observations and statistical data', *Journal of Geophysical Research: Atmospheres*, 111 (2006): *doi.org/10.1029/2006JD007377*.
89. Yinon M. Bar-On, Rob Phillips and Ron Milo, 'The biomass distribution on Earth', *PNAS USA*, 115, 25 (2018), 6506–11.
90. Rachel C. Taylor et al.: *https://www.drylanddevelop.org/uploads/6/1/7/8/61785389/eleventh_international_conference_on_development_of_drylands_2013.pdf* (last accessed: 9 April 2019), p. 69.
91. See *https://www.google.co.uk/search?q=el+nino+years+graph&tbm=isch&tbo=u&source=univ&sa=X&ved=2ahUKEwjmj8m41sDeAhWLalAKHcceD1IQsAR6BAgFEAE&biw=1920&bih=887#imgrc=J9H5x7xgwXyiCM*: (last accessed: 9 April 2019).
92. Global per capita carbon emissions: *https://data.worldbank.org/indicator/en.atm.co2e.pc* (last accessed: 9 April 2019).
93. Thomas Piketty, *Capital in the Twenty-first Century* (Boston: Harvard University Press, 2014).
94. Anthony Atkinson, *Inequality: What can be Done* (Boston: Harvard University Press, 2015).

95. Kate Raworth, *Doughnut Economics: Seven Ways to Think Like a 21st-Century Economist* (London: Random House Business, 2018).
96. Christine Figueres et al., 'Three years to save our climate', *Nature*, 546 (28 June 2017), 593–5.
97. Michael Raupach et al., 'Sharing a Quota of Cumulative Carbon Emissions', *Nature Climate Change*, 4 (2014), 873–9, 873.
98. Naomi Oreskes and Erik Conway, *The Merchants of Doubt* (London: Bloomsbury, 2010).
99. David Coady et al. 'How Large are Global Energy Subsidies', *https://www.imf.org/external/pubs/ft/wp/2015/wp15105.pdf* (last accessed: 9 April 2019).
100. T. Wheeler and J. Von Braun, 'Climate change impact on global food security', *Science*, 342 (2013), 508–13. For a climate change denial/sceptical viewpoint, see *https://www.thegwpf.org/content/uploads/2015/10/benefits1.pdf* (last accessed: 9 April 2019).
101. Josephine Wilson, *Extinctions* (Perth: UWA Publishing, 2016).
102. *Restoring the Quality of our Environment, Report of Environmental Pollution Panel* (Washington: President's Science Advisory Committee, 1965).
103. Simon Gächter and Jonathan Schulz, 'Intrinsic honesty and the prevalence of rule violations across societies', *Nature*, 531 (2016), 496–9.
104. Frederik Obermaier and Bastian Obermayer, *The Panama Papers* (London: Oneworld, 2017).
105. Jonathan Haidt, *The Righteous Mind: Why Good People are Divided by Politics and Religion* (London: Penguin, 2013). See *https://www.goodreads.com/author/quotes/55727.Jonathan_Haidt* (last accessed: 9 April 2019).
106. Emil Durkheim quoted in Haidt [105].
107. Timothy Snyder, *Black Earth: The Holocaust as History and Warning* (London: Vintage, 2016).
108. John Kenneth Galbraith, *The Affluent Society* (London: Pelican, 1971).
109. R. Gareth Wyn Jones, 'Overshooting Limits: Seeking a New Paradigm', in Anna Nicholl and John Osmond (eds), *Wales' Central Organising Principle* (Cardiff: Institute of Welsh Affairs, 2012), pp. 80–123.
110. Prem Sikka: *https://leftfootforward.org/2018/12/prem-sikka-unpicking-the-city-of-londons-new-finance-sector-propaganda* (last accessed: 9 April 2019).
111. Nicholas Shaxson, *Treasure Islands: Tax Havens and the Men who Stole the World* (London: Bodley Head, 2011), and his *The Finance Curse: How Global Finance is Making us all Poorer* (London: Bodley Head, 2018).
112. John Maynard Keynes, *Economic Possibilities for our Grandchildren* (1930), *http://www.econ.yale.edu/smith/econ116a/keynes1.pdf* (last accessed: 9 April 2019).

113. R. Skidelsky and E. Skidelsky, *How Much is Enough? Money and the Good Life* (London: Penguin, 2013).
114. William Bernstein, *A Splendid Exchange: How Trade Shaped the World* (London: Atlantic Books, 2009).
115. Robert U. Ayres and Benjamin Warr, *The Economic Growth Engine: How Energy and Work Drives Material Prosperity* (Cheltenham: Edward Elgar Publishing, 2010).
116. Nicholas Georgescu-Roegen, *The Entropy Law and the Economic Process* (Boston: Harvard University Press, 1971).
117. Ernst Schumacher, *Small is Beautiful: A Study of Economics as if People Mattered* (London: Vintage, 1973).
118. Robert Costanza, Herman Daly et al., *An Introduction to Ecological Economics* (Florida: St. Lucie Press, 1997).
119. Niall Ferguson, *The Ascent of Money: A Financial History of the World* (London: Penguin, 2015).
120. Jared Diamond, *Collapse: How Societies Choose to Fail or Survive* (London: Penguin, 2006).
121. G. Hardin, 'The Tragedy of the Commons', *Science*, 162, 3859 (1968), 1243–8.
122. Oliver James, *Affluenza: How to be Successful and Stay Sane* (London: Vermillion Press, 2007).
123. John Gray, *Heresies: Against Progress and Other Illusions* (London: Granta, 2004), and also his *Gray's Anatomy* (London: Allen Lane, 2009).
124. George A. Akerlof and Robert J. Shiller, *Phishing for Phools: The Economics of Manipulation and Deception* (Princetown: Princetown University Press, 2015).
125. Joseph Schumpeter, *Capitalism, Socialism and Democracy* (New York: Harper and Brothers, 1942).
126. William D. Nordhaus, *Projections and Uncertainties about Climate Change in an Era of Minimal Climate Policies* (New Haven: Cowles Foundation, 2016): *http://cowles.yale.edu/sites/default/files/files/pub/d20/d2057.pdf* (last accessed: 9 April 2019).
127. Arnold Toynbee, *Mankind and Mother Earth* (Oxford: Oxford University Press, 1976).
128. R. Gareth Wyn Jones, *Ynni, Gwaith a Chymhlethdod* (Caerdydd: Y Coleg Cymraeg Cenedlaethol (the National Welsh-language College) and Cymdeithas Ddysgedig Cymru (The Learned Society of Wales), 2017), also 'Ynni, Gwaith a Chymhlethdod', in *O'r Pedwar Gwynt*, 6 (April 2018), 10–13, and 7 (July 2018), 11–13.

Index

I

V

W

Y

Z